ISBN 0 9587584 1 7

Just Talk Pty Ltd, 70 Barada Crescent, Aranda, ACT, Australia

"PHILOSOPHICAL" PREFACE

In 1979 I wrote four "little blue books" in handwriting because they needed a lot of characters that were not available on any typewriter that I had access to. A total of about 100 copies of each were sent to leading scholars or distributed at a conference, but I never received a word in response. In later years I continued the work but made no attempt to communicate it. Long experience had taught me that editors and colleagues were not interested in such unusual material that would puzzle their readers and students. I had no common ground with other academics because I rejected doctrines that did not ring true to me and techniques that did not work in practical affairs.

Now, with time available and access to modern publishing technology, I am making these works more presentable. They provide part of a System of Rational Discourse (SRD), which is a system of logic in English. The first is the basis of the recently published *Definition, Russell's Antinomy, Etc.*. The present work is the second blue book with some amendments and additions. Volume 2, published with the present volume, provides a calculus of aggregates and an interpretation of ZF set theory. Subsequent volumes will include the remaining two blue books interspersed with further works that examine extant logical doctrines in the light of SRD. Volume 3, in preparation, will give a *practical* treatment of numbers and arithmetic and a critical look at probability theory. Later works will deal with predications, propositions and modalities, and provide a rigorous treatment of English infix quantifiers, 'every', etc., showing how they but not Frege's prefix quantifiers accord with traditional syllogistic. Along the way, various standard logical puzzles like induction, the formulation of scientific laws, counterfactuals, fallacies and incoherencies of statistics, techniques of quotation, various paradoxes, etc., will find non-puzzling treatments.

Among the desiderata of a system of rational *discourse* are precise specification of meaning, public acknowledgement of meaning, rigour of proof of *doctrine* (not provided by mere mechanical certification of notation) and, in application, truth of assertions. There are philosophical arguments that each of these is impossible of total achievement. I make no prior appraisal of such arguments since in doing so I would be indulging in the undisciplined philosophizing that SRD is intended to replace. Because SRD is rigorous, I avoid extensive traditional intuitive philosophical discussion in the body of the work but provide this preface to introduce the reader to the unfamiliar enterprise. Any doctrines expounded in SRD, whether purely logical or applied, are

1) of precisely specified meaning, according to explicit definitions,

2) of publicly acknowledged meaning *to anyone who heeds the definitions*,

3) rigorously proved, and checkable by rules for dealing with notations, and

4) true, all but 3), which is absolute,

to a degree not less than that initially entertained by discussants.

No reliable system can be creative and produce these desiderata from nothing, but a good system can be conservative and not impair them. SRD is, in an old-fashioned sense, a system of Logic. But modern Logic has developed as a treatment of multiple substitutes for natural language, not as a systematization of it. These substitutes are typically topics for metadiscussion (ultimately in unformalized natural language) rather than means of talking about the world. My task is to provide such a means, which is why I emphasize discourse.

A consequence of my economical method is that I avoid long explanatory disquisitions and make no explicit references to the literature, with the result that my work was once unkindly said to verge on plagiarism. It is now worth while to forestall further such criticism. There are certain notions and problems that are part of the furniture of a scholarly discipline, the provenance of which it is known to trained workers. I refer to such matters without the apparatus of bibliographical references because these volumes are not textbooks for beginning students but are addressed to experts. But I make no claim to originality for *anything* in these works. If anyone claims precedence on his own or someone else's behalf I leave adjudication to others.

The motivation for the whole project seems to be purely idiosyncratic with me. I seem to be alone in seeking a means of *proving* what I *say*. Right away, most people will be puzzled by that. But some people cannot and some will not talk plainly. I repeat: I seek (and provide) a means of *proving* what I *say*, a means that anyone else can use if motivated to. Some may already doubt my veracity, my competence or even my sanity. Don't *they*, when they want to, conduct rigorous proofs in natural language? No, they do not. More generally, has there not been a couple of millennia of work in Logic, seeking and providing what I claim to seek? No, there has not.

Some academics use only natural language without any rigour at all, producing what is unkindly called waffle. All will agree with that, but differ in picking out the wafflers. Others use rigorous systems of notation and do not care about any interpretation. They derive *graphical* results, but since they do not *say* anything they have nothing to *prove*. Of course they will claim to (and actually *do*) "prove" (i.e. derive notations according to rules), but that claim is metaphorical or ambiguous, not plain talk. Others use rigorous systems that have no canonical (definitive, official) interpretations, but try to make their meaning public by means of *informal glosses* in natural language. These glosses are different from author to author and from occasion to occasion and cannot be rigorously connected with the graphical results they are supposed to match. They have no formal value at all. In important cases the glosses not only do not match the formal notations but *utterly and bizarrely misinterpret* them. I have given a wide-ranging account of the chaotic academic scene that I observed from the inside for three decades, and its bearing on people in general, in *The Emperors' New Logic*, a non-technical work for the general reader. Whatever your leanings, reading that should give you a good idea of what I am seeking, my motivation, and how my experience led me to the start the venture. You will then better appreciate the present work.

My twofold aim is to provide a means of rigorous natural-language proof and to provide the missing correct interpretations for standard systems. And I find myself in a predicament in trying to get myself understood. I present everything needed for clear understanding, with all due explanations and rules. But I see myself *apparently* open to the charge of needless difficulty and obscurity. One may find some things that I say worth considering, but on the whole my discourse falls into one of three categories of things that *immediately seem* obscure or not worth saying.

1) Some things I say are *obvious* and *trivial*. The reason why I need to say and repeat them is that though people think them obvious, and feel that I labour a point too much, in their own discourse they flagrantly disregard what is obvious and talk in strange ways about strange subject matter in disregard of the obvious. I need to keep on reminding them of what they have already acknowledged. A common reason for people's wavering is that they initially use words literally, locally and precisely, then use them ever more metaphorically, globally and vaguely. Then what they later say may be true if what they mean can be pinned down, though irrelevant to the initial discussion, but false in the original literal sense. In SRD there is no use of metaphor, so such freedom of interpretation is illicit. A common case is to acknowledge the literal meaning of 'language' (or 'tongue') as dealing with ways of talking, like English, Latin, Hebrew, etc., i.e. with natural languages, then later talk about the "language" of flowers, the "language" of bees, "body language", the "language" of mathematics and a host of other things. All of these "languages" are means of expression, each with its own characteristics, but a language is a particular, familiar means of expression with *its* own characteristics, and that is what discussion was initially concerned with. The "languages" of logicians mimic (or distort) the structure rather than the meanings of utterances. Those of linguists are "sets" of sentences. For such reasons, the words 'language', 'information','encode' and others of the "information-processing" school are so corrupted that nothing definite can be said using them. The rampant wild metaphors fool the public and, in many cases, the authors. It would come as a great surprise to Mendel and Darwin that there is a genetic *code*, and to Sherrington that nerve cells nowadays *talk* to one another. Such careless or wilfully misleading doctrine has no place in any serious intellectual enterprise.

2) A prime feature of a skill is that it is applied spontaneously, without deliberation. Some skills are learned by practice. Others, like walking and talking, are picked up on the basis of native endowment and suitable experience. All but the most unfortunate people can walk and talk, most of them having no more idea of how they do it than of how they digest their food. It doesn't matter. Use of these skills is like the visible part of a skyscraper, but the principles of them are like the infrastructure, the underground rooms that provide essential services but are invisible to an ordinary user of the building. People who use a skyscraper can be very surprised if taken on a tour of the rooms below the ground. They find objects and processes that they had never dreamed of. What they are shown is unfamiliar and complicated, and hard to understand without technical training. What is important is not whether ordinary people understand what happens underground, but whether the infrastructure of a building effectively but unobtrusively supports the use of the skills that office workers, shop assistants, cleaners, managers, householders, etc. have. Such use *depends* on matters of which many skill users are justifiably totally ignorant.

A *proper* analysis of an intuitive skill like walking, talking or proving is like a tour of the infrastructure of a building. What is important is not whether ordinary people understand the account (they will not, without training) but whether the technical account accurately reflects the skill in action.

This leads to my second suspicious way of talking. My task is to provide an effective account of the bases of various logical and linguistic skills that ordinary people put to good use on occasion but sometimes unknowingly misuse. *Of course* my account of the infrastructure is hard to understand. If it were otherwise there would be no problem in the first place; everyone would already know all the details. What matters is: if someone abandons intuition and follows my unfamiliar rules, does he wind up thinking and talking as he did intuitively in cases where he gets correct results and does he avoid puzzlement and incorrect results that he cannot avoid intuitively?

3) At other times I pick out some seemingly weird, barely intelligible topic and examine it under a linguistic and logical microscope. So people will think "Why does he use this sledgehammer to crack such peanuts?". I explain why.

A bone-by-bone and muscle-by-muscle theory of walking that accurately specifies a silly walk by John Cleese, no matter what its virtues and value, is not a theory of ordinary walking. If you follow the rules you will be doing a silly walk, not walking. A purported theory of an intuitive skill that specifies some unfamiliar end procedure that no ordinary person ever engages in is not a theory of that skill but a theory of a silly-walk substitute for it. For example, a theory of language use that has a speaker or homunculus going through an elaborate sequence of constructive grammatical steps, and has a listener waiting till the end of a sentence then going through elaborate analytic grammatical steps is a silly-walk theory of language use. It may be an accurate theory of a grammar lesson, where students follow the teacher's instructions, but it is nothing *provably* more. Though Chomsky has repeatedly stressed that a *generative* grammar is just a *rigorously presented* grammar (of whatever linguistic character), he has not prevented badly trained followers from thinking it is a theory of speech *production*. Thus is a major enterprise compromised by unfortunate choice of a title.

I do not present elaborate analyses of trivialities and the fantasies based on them either for my own amusement or to "educate" a reader. I do so because these trivialities and fantasies are the proper interpretations of extant logical and mathematical systems, which are silly-walk theories. This appraisal is not generally known, implausible on the face of it, and unwelcome when seen. If I merely made such charges and presented my material in ordinary language, without the technicalities and the disciplined language, my remarks could be rejected out of hand by experts and regarded as my misinterpretations by others. I need to jump through the technical hoops in order to demonstrate that my ordinary-language conclusions are correct.

The whole (wrong) impression taken from these methods can be that it is ironic that this self-proclaimed champion of practicality, rigour and meaning, and of the value of common sense and ordinary language in achieving them, should talk so unfamiliarly about some things that are hard to understand and other things that are of no conceivable use and are actually obstructions to useful and worthwhile discussion. I *agree* that some are both hard to understand and useless. But I must elaborate them in unfamiliar ways to *make quite certain* that this is so, in the most rigorous fashion.

But my work is not purely critical. As well as exposing inadequacies, I provide formulations of doctrine that are more intelligible than those I criticize. Indeed, some of the criticisms could not be made until proper accounts were available. SRD is unlike any conventional logical system in motivation, starting point, mode of presentation, content and destination. But it does not ignore technical developments of the last hundred years. To do that would suggest the obviously false claim that there is nothing of logical value in these developments. Frege's contribution, in particular, (in its modern format) though *not as sound or as versatile* as is generally supposed, is, with some modification and extension, an essential part of SRD.

Frege's work introduced so much novelty of content and so much new rigour that it made little impression at the time on logicians and mathematicians. But it is flawed because Frege was a mathematician, with mathematical interests and goals, and an unexamined adherence to the mathematical way of proceeding. This way, divorced from natural language and from the everyday world, is itself a source of logical problems that are insoluble in such a context, as the history of 20th-century logic and mathematics attests. *Definition*, ... gave an example of avoidance of rather than ineffective coping with problems, this volume gives an example in zulogy, and Volume 2 gives examples in its treatment of aggregates and ZF set theory.

A presentation of SRD from its absolute foundation would now be similarly novel and intimidating. Such must be reserved for an occasion when the content is familiar, and the reader is led in a continuous path over individually familiar tracts of territory. For now, I rely as much as possible on conventional material and methods, explaining innovations piecemeal. So here I jump right in, assuming sentential calculus and lower (first-order) predicate calculus (LPC), though the latter is amplified here and the former will in a later work be clarified, not in content but in modality. Some later work, adumbrated in my 1979 works, will provide even deeper yet more practical foundations for what I present here. The apparent incompatibility of greater logical depth and more practical basis is an illusion. If work is started in the conventional way, outside language, then the quest for foundations leads further and further away from language and ordinariness. But SRD, starting in English, finds its basis in ordinary language and in the ordinary world, *in the most practical way possible.* I deliberately express this cryptically to enhance the reader's astonishment when he finally sees how it is done.

A discourse is a self-contained domain of discussion, whose author may elect it to be subject to logic or not. Either way, there is no possibility of inconsistency across discourse boundaries. *Oliver Twist*, *Lord Of The Rings*, and *Through The Looking-Glass* are segregated unregimented *fictional* discourses. It is silly to contradict them, and no one ever tries, because an author of fiction *cannot be wrong*; what he says is right just because he says it. To try to alter fiction is to become an author of fiction yourself, in which *you* cannot be wrong. Since there are multiple incompatible regimented *axiomatic* systems in logic and mathematics, they must be segregated. Those whose authors purport to talk are discourses, and have the defining character of fiction in that their doctrines are right by authorial say-so, and cannot be wrong.

By universal agreement among scientists, Newton's scientific works are in a common discourse with Einstein's so there is inconsistency, and Newton's doctrine must be disavowed in treating relativistic phenomena. But there are currently only informal, intuitive connections between scientific notation and English. SRD, as it gradually develops, is offered, not as a doctrine in scientific discourse, but as the common regimented English dialect for it.

The unavoidable end result of all these novel features is that these volumes require hard work from a reader. As Euclid said to a king who asked for an easy treatment of geometry, "There is no royal road.". The hard work is both *technical*, which logical experts will cope with, and *linguistic*, which some non-technicians will cope with. But a single side of the coin will not suffice. So most lone readers, no matter what their intellectual backgrounds, will feel they are fish out of water. The work is best studied by groups with access to both sorts of expertise and with conflicting prejudices. Without such cooperation and disagreement, individual students can easily accept or dismiss SRD for jointly incompatible reasons.

If readers skip the "boring" (rigorous) bits and read the "interesting and possibly important" (literary) bits, they will all come away with their own interpretations, and miss the essential point of SRD. In SRD I discipline the use of a familiar word if one of the family of its meanings is near what I want. If a word, like 'symbol' or 'representation' (in general, any widely used cognitive word), is incurably debased with various unclear meanings then I avoid it and invent a neologism. Everyone realizes that a neologism is no good unless its meaning is well specified, and no one would expect to understand what I say using one, without consulting my definition. What one has to realize here is that the same goes for familiar words. But to understand my definitions one must know the general principles of defining, and to know that one must study the general framework of SRD.

The nearest thing to a "message" in these volumes is that ordinary talk and common sense, though not immediately rigorous, are the best basis available for the development of rigorous exposition. They deserve regimentation, not rejection. Because there are too many cooks, these admirable tools get used in various ways, not all of them appropriate. But among these various spontaneous, accidental or deliberate ways of talking there are some intellectually useful ways, to find and systematize which is my self-appointed task, whose outcome is SRD.

I expect that the whole work will at first appeal to very few workers, but new students, without commitment to extant systems, will naturally gravitate towards a simpler, more versatile, more familiar logic. The two intellectual methods calling on precisely manipulated meaningless notations and clumsily used meaningful sentences respectively are entrenched in the intellectual mindset, even among those who have never thought about the matter. People are drawn to one side or the other by temperament or training. I suspect that the aim of precisely used sentences, which drives SRD, has never occurred to most people, has seemed already achieved to the imperceptive, and has seemed unachievable to the few others. Now that the aim has been proclaimed, and is being achieved step by step, the outcome will be unwelcome to all but those who have or are prepared to develop skill in both fields.

It is well known that amateurs who venture into logic or mathematics use notations imprecisely and derive unwarranted conclusions, and it is pretty obvious why. Expertise in such matters demands a certain aptitude and much training and practice. People might be expected to have sufficient self-awareness to realize when they are out of their depth and leave it to charlatans and fools to flaunt their incompetence. But, sadly, it is not so, in part because all but experts are taught sanitized logical doctrines that have no warrant at the highest level of expertise. It is quite common for people to have a doctrine expressed in English then, in an attempt to impress, give a garbled account of it in formal notation. All "practical applications" of set theory illustrate this. Though experts are, of course, aware of the fact, they are too polite to say so. But intellectual endeavour is too important a matter to be retarded by politeness.

It is not so well known that the converse also applies, though it should be obvious. These volumes will show again and again, at all levels of discussion, that verbal matters, just like notational ones, have their complexities. Expertise demands a certain aptitude and much study, but there is no reliable source of training. Traditional grammarians and philologists developed their disciplines with as little concern for language *in use for communication* as do logicians and mathematicians. And modern linguists have adopted the techniques of logicians so they can be put in the same category.

Many scholars are aware of the damage done to discourse by ambiguities and confusions in the use of lexical words. Like the injuries, exposures, exertions, poisonings, etc. that disturb bodily health, they are fairly easy to detect and guard against. So my tirade in 1) (p.v), whether accepted or rejected, is an example of a familiar tactic. In contrast, a more basic trouble is the multiple uses of grammatical words like 'and', 'of', 'that', 'have', 'be', 'not', 'the', etc. Like microbiological agents such as bacteria, viruses and genes, these wreak their havoc invisibly and unsuspectedly before extensive technical investigation exposes them, then after detection, further technical work is needed to control them. It is my interest in these technical activities, treating the *meanings* of grammatical words, that makes my work as unfamiliar today as Pasteur's was 150 years ago. People cannot see what the fuss is about. But ignorance or disregard of verbal subtleties of meaning leads to errors of verbally expressed doctrine, *even without* notational mistakes in axiomatic systems.

The rantings of fanatics and the woolly-minded ravings of New-Age theorists are taken with a pinch of salt by sober, practical people. Unless they become the driving force for repressive governments or antisocial cults they serve a harmless purpose of entertaining the gullible in private. Likewise, the garbled notations of technical incompetents carry no intellectual weight, and are influential only among fellows of the authors. Many of these authors and their fellows are academics, whose products take on a deceptive patina of precision, especially evident in linguistics and "cognitive science".

But the garbled verbal output of skilled technicians, unlike the garbled technical output of non-technicians, carries a great deal of intellectual weight, with widespread bad academic and social consequences, because it comes from authors who are respected because of their technical skill. Perceptive but untrained sceptics are blinded with science. Both experts and those influenced by them naively suppose that experts are able to tell the public what their results mean. They are wrong in this. Some of their results, as experts see them, are so repugnant to common sense that experts tell "little white" lies (but lies, none the less) in public communications, including textbooks. Those lied to become unwitting, incompetent purveyors of such lies, and the ultimate recipients, school children, are given intellectually and practically inferior tuition, of which the now moribund or defunct "New Maths" of the sixties is the most ballyhooed example.

It would be naive to suppose that there is a conspiracy here. The originators of bad doctrines are acting in good faith and do not know why they are being criticized, what their fault is supposed to be. Since their technical proceedings are virtually free of criticism, they see no reason to be under attack. Their only response can be that the attack is philosophical and therefore irrelevant since their work is not philosophy. That defence is perfectly in order, given one condition: that they confine themselves to moves in their private games *without the verbal running commentary*, or else learn to use language properly so they can give an *accurate* commentary.

The differences between conventional systems and SRD are not mere matters of detail. SRD contains standard Logic as a reinterpreted minor but *working* part. It contains set theory as a reinterpreted minuscule and *idle* part, whose verbalization is like the impressive, thunderous, exhaust roar of a flashy but inefficient sports car. Standard tasks, when worthwhile at all, are better done SRD's way, the way of disciplined ordinary language. But the main problems of standard Logic, and *a fortiori* of set theory, are generated by inbuilt limitations, and their "solutions" only generate more problems. These problems do not find any solutions in SRD; they simply do not arise, and are not missed.

I have here made strong "philosophical" criticisms of academics and strong claims for SRD. Now, as devil's advocate, I say: of themselves, they carry *no weight whatsoever*. They are not even worth rebutting in kind for, by the same token, such rebuttals carry no weight whatsoever and are not worth making. The purpose of this preface is merely to let the reader know what the whole project is about. What will give weight to the criticisms and the claims is the rigorous verbal proof that SRD offers. Only similarly *rigorous verbal proof* will give any weight to attempts at rebuttal.

CONTENTS

Rigour without meaning is empty.
Meaning without rigour is vague.

INTRODUCTION

In this volume (Volume 1, though not the *logical* start of the work) I present the basics of a System of Rational Discourse (SRD): a rigorous, neutral way of talking for use when one is concerned with precise expression and sound argument. SRD is a way of talking because it is a portion of English. It can be extended indefinitely to include more and more locutions, at the price of eliminating their ambiguity and vagueness. Though these features are tolerable, useful or welcome in both daily life and literature, they are an impediment to rigour. SRD fulfils its purpose by being disciplined according to explicit rules. It is neutral because it allows a user

a) to discuss any kinds of subject matters concurrently, since it neither imposes nor invites any prior specification of a "universe of discourse",

b) to say what he likes consistently about subject matter, since its concern is rationality, not honesty or omniscience (others will test an author's veracity),

c) to use, as a supplement to what is provided, any properly introduced dictionary expression he chooses,

d) to provide any properly formulated definition for a new expression.

SRD is a designed *English dialect*, not a doctrine whose truth or falsity one might wish to determine; it can be *used* to present doctrines. This feature, its motivation and its benefits, will be explained in a later volume. On all aspects of what is consistently said, SRD is permissive, leaving it to users to justify their unproved judgments by appeal to some subject matter. On important aspects of how a judgment is expressed it is as permissive as precision allows. But if users "improve" the dialect, their subdialects will become non-standard and impair public communication. The trouble caused by different versions of computer programming methods is a warning against introduction of idiosyncratic changes.

SRD is from the start presented as a meaningful (using a redundancy for emphasis) language being used for communication, not as an uninterpreted notational system (a "language") ultimately described in unregimented ordinary English and awaiting a "semantic" assignment of meaning. Every notation of the pure calculus is given a definite *grammatical* English reading and is treated merely as a graphical and visual, not vocal or aural, abbreviation of that reading. Interpretation of primitive expressions is perforce intuitive, calling on a user's prior understanding of English, but there are few primitives and the systematic way they are used guides different readers towards a uniform precise interpretation. As for any system of such a design, perfect grasp of primitives is a practical matter, not a logical one. The meaning of a defined expression is determined by those of the expressions used in the definition.

Given that modern Logic is mainly concerned with truth, hence with assertions, as a route towards meaning, my stand is novel. But the conventional stand, rather than mine, is novel in the light of ordinary langauge and common sense. Meaning must be grasped before truth can be judged. As John Doe might say "How can I judge whether what you say is true or false without knowing what it is you say? How is this difference brought about, why has standard Logic not anticipated SRD and why can standard Logic not follow SRD? Because of zulogy, which is a formal treatment of ontology.

The exposition shows that I pay the price for my non-standard approach, in my long journey to reach apparently easy goals, but that is a price well worth paying. In computing parlance, SRD is *robust* in the face of problems that cause trouble for standard Logic.

Since these works are written for trained persons, I do not refer to theorems of LPC in annotations, taking it for granted that readers know the relevent items or can look them up.

PREDICATE CALCULUS

Basis

A sentential calculus is taken for granted here, employing signs '&' for 'and', 'v' for 'or', '~' for 'it is not the case that', '... ⊃ ---' for 'if ... then ---', and '≡' for 'just if'. '~' may be spoken briefly as 'non', not as 'not'. Though the new calculus is compatible with any of a variety of calculi it is most conveniently treated for a start in the context of the standard "two-valued" calculus of *Principia Mathematica*.

Notation

Atomic formulas of the new predicate calculus employ the word order of English with an explicit copula, not Frege's "functional" notation '*f*(*x*)', '*f*(*x*,*y*)', etc.. 'Socrates is mortal' is an example of the form '*x* is *f*'. What occurs before 'is' is a putative, not necessarily genuine, referential expression, most simply a proper name. What occurs after 'is' is a putative descriptive expression, most simply an adjective or a simple indefinite noun phrase, e.g. 'a table'.

The copula is optionally omissible in relational formulas so 'Russell is a colleague of Whitehead' may be graphically abbreviated to 'Russell coll Whitehead'. Except when spoken abbreviations are explicitly given, as 'non' for 'it is not the case that', graphical abbreviations are spoken *in full English.* The example has the form of '*x r y*', spoken '*x* is *r* of *y*'.

If you are not willing to speak in full English you are wasting your time here.

Blankmarkers

Since the role of a phrase is determined by its position relative to the copula or to some other logical constant, the calculus does not need conventional variables or schematic letters having predetermined ranges of substituends or replacements. Letters '*p*', '*x*', '*f*', '*r*', and their respective coordinates are written in systematically different places for perspicuity, and as far as possible conventionally. But each is simply a cross-referencing blankmarker indicating blanks that may be uniformly filled by any notation not explicitly exempted — other blankmarkers, sentences, names, interpreted predicates, non-phrases, vacuous names or verbal garbage. Thus '*p* is *r*', '*g* is *q*', '*f* is *x*', 'a 4b+? is con', 'blue a the is dill', etc. are all feasible atomic formulas. Such freedom of notation given a user contributes to the desired philosophical neutrality.

Zulogy

In the absence of censorship of notation, the calculus provides minimal internal discipline, sufficient merely to ensure that freedom of blank filling does not produce gibberish, i.e. garbage posing as meaningful utterance; garbage not so posing is innocuous. The method of controlling notation depends on zulogy, which is a formal, not material, ontology. To forestall profitless, presystematic, metaphysical squabbles I introduce zulogy as the discipline concerned with a certain notational constant '*Z*', pronounced "zu". This notation is primitive and absolutely nothing is intuitively suggested at present regarding its interpretation. Its total significance is specified by this axiom.

A1.1 $x \text{ is} f \supset x Z$

By substitution of '$r\ y$' for 'f', (a procedure permissible because 'f' is just a blankmarker, not a "one-place predicate letter") and elision of 'is',

1.1 $x\ r\ y \supset x Z$

Since 'y' here is just a blankmarker, not an "individual variable", 'r' is not a "two-place predicate letter". Before proceeding, make sure you are speaking this theorem properly, and ponder why I need to urge people to do just what is natural.

According to A1.1, anything said about x warrants that $x Z$, so to say that $x Z$ is to make the most general, least informative characterization of x that the calculus permits. Because 'is' is merely a sign for predication, and 'f' as a blankmarker imposes no restriction, it is the most general characterization expressible at all. As 1.2 shows, if that characterization is not applicable then none is.

1.2 $\sim x Z \supset \sim x \text{ is} f$ [A1.1]

Thus the nature of zulogy is illustrated as follows. If each of a number of users imports his own most general, honorific, ontological, epistemological, phenomenological, mystical, astral or ...-ical characterization to the effect that x, e.g., is, be, exists, subsists, is a thing, is immanent, is transcendent, is a spatio-temporal object, is a 4-D point, is perceived, is here-now, is thinkable, is an element, is a class, or whatever, then, if x is so characterized then x zu; if $\sim x Z$ then $\sim x$ is so characterized.

If x is "not a thing" then x zu; if $\sim x$ zu then $\sim x$ is "not a thing", and it does not follow that x "is a thing". Likewise for every other specimen above. No matter what disputes metaphysicians may have about the the relative breadths of their ontologies, because of its purely formal concern zulogy always plays the ontological ace of trumps.

More sweepingly:– if $\sim x$ zu then $\sim x$ is blue, $\sim x$ is a philosopher, $\sim x$ is a unicorn, $\sim x$ is L3!zble. It adds nothing to say (informally) that $\sim x$ is (whatever you like) for this is just the import of '$\sim x$ is f'. Although the calculus is master of formal zulogy a Robinson Crusoe is master of his material zulogy. Socially, users in dispute will argue about material zulogical claims; using SRD, they will argue soundly.

A constituent of The Universe

Nothing in the calculus as so far presented here provides that anything zu, and it would be contrary to the spirit of sincere, straightfaced, plain talk, without pretence or pretension, to claim axiomatically that something zu. On pretence: although anyone is at liberty to play games, the present serious attitude precludes make-believe; on pretension: although God said "Let there be light." and there *was* light, I have no confidence that *fiats* of my own would be as effective. But despite this impartial intent the very character of a more primitive version of the calculus, presented in a later volume, ineluctably provides a zulogical theorem that must be acknowledged here as a temporary axiom, to prevent inconsistent extensions being added.

TA $J Z$

The nature of J is not relevant here since TA is used here only to establish that something or other zu.

Quantification

Open quantification is formally like standard "universal" quantification.

A1.2 $(x)(\phi(x)) \supset \phi(y)$
if, whatever x is, $\phi(x)$ ...

The standard ungrammatical reading "for all x' is rejected. Because of the prefix '(x)', 'x' in A1.2 is not a blankmarker, but a pronoun. There are not enough pronouns in English for the needs of SRD (which needs indefinitely many, as LPC needs "variables"), so such letter pronouns can be pronounced as they are spelled. Because 'y', as a blankmarker, takes any filling, open quantification gives a user *carte blanche*. There is, of course, only one universe, whatever disputes there may be about its composition, and besides that nothing. So open quantification cannot be supra-universal zulogically, but regarding notation it provides a supra-universal licence, as does, of course, a detached open formula with blankmarkers.

The converse of A1.2 is not a theorem but the usual rule applies.

R1.1 The result of prefixing any detached formula with an open quantifier for any of its blankmarkers is detachable.

Zulogical quantification (cf. standard "existential" quantification) is restrictive. Having zulogical commitment, it imposes a zulogical condition.

A1.3 $(\#u)(\phi(u)) \equiv \sim(u)\sim (u\ Z\ \&\ \phi(u))$
u is such that $\phi(u)$ ...

Because of the prefix '$(\#u)$', 'u' is here a pronoun, not a blankmarker. The novel sign '#' distinguishes zulogical quantification from standard "existential" quantification, associated with '∃'. The latter notation does not have zulogical commitment and is not used in SRD. From A1.3,

a $\sim(\#u)(\phi(u)) \equiv (u)(\sim u\ Z \vee \sim \phi(u))$

b $\sim(\#u)(\sim \phi(u)) \equiv (u)(\sim u\ Z \vee \phi(u))$

c $\sim(\#u)\sim \phi(u) \supset \sim y\ Z \vee \phi(y)$ [A1.2]

d $y\ Z\ \&\ \sim \phi(y) \supset (\#u)\sim \phi(u)$

1.3 $y\ Z\ \&\ \phi(y) \supset (\#u)\ \phi(u)$

a y is $f \supset y\ Z\ \&\ y$ is f [1.1]

1.4 y is $f \supset (\#u)\ u$ is f [1.3]

* $\sim y$ is $f \supset (\#u)\sim u$ is f

It is the availability of the last case, with a negated antecedent and an affirmative consequent, that rules out an axiom to the effect that $(\#u)(\phi(u)) \equiv \sim(u)\sim \phi(y)$. In the absence of this analogue of the standard axiom (or equivalent "definition"), A1.2 does not itself provide 1.5, but 1.5 is still a theorem for another reason.

1.5 $(x)\ \phi(x) \supset (\#u)\ \phi(u)$

Instead, by A1.2 and 1.3,

a $y\,Z \,\&\, (x)(\phi(x)) \supset y\,Z \,\&\, \phi\,(y)$

b $\supset (\#u)\ \phi(u)$

c $y\,Z \supset {\sim}\,(x)(\phi(x)) \vee (\#u)\ \phi(u)$

But because TA satisfies the antecedent as a theorem, the calculus as a whole provides 1.5 from *c*. By 1.5,

a $(x)\ x\,Z \supset (\#u)\ u\,Z$

The antecedent, which by A1.2 would provide gibberish, such as '*?A7Q *Z*', is not a theorem. But by 1.3,

a $\mathrm{J}\,Z \supset (\#u)\ u\,Z$ [1.3]

1.6 $(\#u)\ u\,Z$ [*a*, TA]

On the grounds that J, some determinate thing, zu, the calculus acknowledges that something is a constitutent of the universe, for this is the import of zulogy.

In place of A1.3, standard LPC has the following, without the zulogical condition.

* $(\exists u)(\phi(u)) \equiv {\sim}\,(u){\sim}\ \phi(u)$

* $\phi(y) \supset (\exists u)\ \phi(u)$

* $f(y) \vee {\sim} f(y) \supset (\exists u)(f(u) \vee {\sim} f(u))$

* $(\exists u)(f(u) \vee {\sim} f(u))$

The foregoing discussion shows how, with "universal" quantification having no "existential" import, being compatible with an "empty universe", LPC itself does not confer such import on '∃'. On the strength of an instance of the tautology '$p \vee {\sim}\, p$' (expressing *no affirmation*) it claims some unnamed, uncharacterized *je ne sais quoi* for its "universe of discourse", without the evidence of any such thing as J.

This unwarranted claim stems from a presystematic decision to treat '∃' "existentially" coupled with a failure to design LPC accordingly. There is nothing *formally wrong* with the '∃' quantifier or its treatment as a notational abbreviation. What are wrong are calling it existential, with unwarranted ontological implications, pronouncing it "there exists ...", making the claim explicit, and the ungrammatical, incoherent 'there exists an *x* such that ... *x* ...' where '*x*' takes on two incompatible grammatical roles. In contrast, the corresponding presystematic decision to treat '#' zulogically is supported by the design of A1.3. Without J the new calculus could not claim as a theorem that something zu, but it would still permit any user discussing his subject matter to derive a corresponding conclusion in an application of SRD.

Frege's logic is *formally* correct, as far as it goes, and his quantifiers are undoubtedly, as widely acknowledged, the greatest contribution to Logic since Aristotle. But his arithmetical interests and mindset deterred him from investigating language in detail. His concern was explicitly with "concept writing", not language. Modern workers, however, attempt to use it to analyse language, but they have not elaborated his procedures to produce a more versatile calculus, as I do in SRD.

ABSTRACTION, IDENTITY AND DEFINITION

Abstraction

The previous work *Definition, ...* presented and explained an abstraction procedure with a reading and rules. It was written without excessive novelty, first to make it as accessible as possible to a conventionally trained reader, and second to show that the avoidance of Russell's antinomy does not depend on a particular calculus, such as that now being presented. But the wider goals of the present work, in particular the introduction of zulogy, now demand elaboration of the earlier principles.

The open abstraction schema '[ϕ(it)]' is pronounced "being such that phi it". With suitably introduced notation, an abstract is a referential sign that refers to an attribute. '*E*' is a primitive predicate constant making a two-place formula and is pronounced "an exemplifier of" or (with the implicit copula) "exemplifies".

An abstract is introduced by the abstraction axiom schema A2.1, where '*x*' is not free in the abstract, i.e., no blank inside an abstract is linked to any blank outside it.

A2.1 $x\,E\,[\phi(\text{it})] \equiv x\,Z\,\&\,\phi(x)$
x exemplifies being such that phi it

Previously '*x Z*' was omitted. The special pronoun 'it' refers to the referent of a filling for the blankmarker '*x*'. It requires A2.1 to be subject to rules given and explained in *Definition, ...* . R2.5 is new, and has no application yet since the matter of modality has not arisen. Modality will be treated when the fundamentals of SRD are presented.

R2.1 An abstract permits substitution for any blankmarkers it contains

R2.2 In treating any instance of A2.1, whatever notation replaces the subject blankmarker (that preceding '*E*') must replace *just every non-embedded occurrence* of 'it' in that instance.

R2.3 A pronoun is not a substituend.

R2.4 An abstract may be assigned a constant coreferring expression in a definition only when blankmarkers are equally present in both expressions, and 'it' occurs in only a single place in any atomic formula.

R2.5 No modal notation is to be put in an abstract.

The ontological ace of trumps mentioned above is [it *Z*]. By A2.1,

2.1 $x\,E\,[\text{it}\,Z] \equiv x\,Z\,\&\,x\,Z$

a $x\,E\,[\sim \text{it}\,Z] \equiv x\,Z\,\&\,\sim x\,Z$

2.2 $(u) \sim u\,E\,[\sim \text{it}\,Z]$ [*a*, R1.1]

[~ it *Z*] is an attribute that nothing can exemplify.

Identity

Despite much controversy about means (and about meaning) identity is conventionally characterized as reflexive, symmetrical and transitive, and as ground for unrestricted interchangeability of equireferential notations.

R $x = x$

S	$x = y \supset y = x$
T	$x = y \;\&\; y = z \supset x = z$
I	$x = y \supset (\phi(x) \equiv \phi(y))$

The first item cannot be tolerated as a theorem here since it demands censorship of notation. If '*x*' is a blankmarker, as here, it swamps discourse with endless gibberish such as 'abcd = abcd'. Then it overpopulates the universe with chimaeras since, by 1.1, abcd zu.

A generic identity attribute $B = y$ cannot yet be defined since there is no procedure for making definitions, and identity itself is involved in any such procedure. So I introduce a generic identity predicate ' = *y*' in an axiom, with a zulogical condition. Note the ordinary quantification employing a notation '*u*' in a referential place, a place given the appropriate effect by '*E*'. This grammatical, intelligible procedure in SRD (and ordinary English) achieves what incoherent "quantification" employing '(ϕ)' fails to do in PM. Others have diagnosed this problem and devised various solutions.

A2.2 $x = y \equiv x\,Z \;\&\; (u)(x\,E\,u \equiv y\,E\,u)$
x is (identical with) *y*

'identical with' can be elided, since the identity sign '=', when properly used, is always followed by a referential notation. Even though English does not employ this sign, the presence of a reference sufficiently indicates the meaning. (A *reference* is a referential notation, not necessarily having a *referent*, what the notation purports to refer to. 'Pegasus' is a reference but there is no such referent as Pegasus.) So users of ordinary English, without being aware of any problem, are not wrong in commonly making the elision. If one bears this in mind there is no danger of confusing the 'is' of predication with the 'is' of identity, and I shall generally use the brief expression in verbalizing SRD notation.

a	$x = x \equiv x\,Z \;\&\; (u)(x\,E\,u \equiv x\,E\,u)$	[A2.2]
2.3	$x = x \equiv x\,Z$	

This is principle **R**, with a zulogical condition, and a zulogical consequence.

2.4	$x\,Z \supset (x = y \supset (u)(x\,E\,u \equiv y\,E\,u))$	[A2.2]
a	$x = y \supset (x = y \supset (u)(x\,E\,u \equiv y\,E\,u))$	[1.1, 2.4]
2.5	$x = y \supset (u)(x\,E\,u \equiv y\,E\,u)$	[*a*]
a	$x = y \supset (x\,E\,[\phi(\text{it})] \equiv y\,E\,[\phi(\text{it})])$	
b	$x = y \supset (x\,Z \;\&\; \phi(x) \equiv y\,Z \;\&\; \phi(y)))$	[A2.1]
c	$x = y \supset x\,Z \;\&\; y\,Z$	[1.1, *b*]
2.6	$x = y \supset (\phi(x) \equiv \phi(y))$	[*b*, *c*]

This is principle **I**.

a	$x = y \supset (x = x \equiv y = x)$	[2.6]
2.7	$x = y \equiv y = x$	[2.3, 1.1]

This is principle **S**.

a	$x = y \,\&\, y = z \supset z = y$	[2.7]
b	$x = y \,\&\, y = z \supset zZ$	[*a*, 1.1]
c	$x = y \,\&\, y = z \supset ((u)(x\,E\,u \equiv y\,E\,u) \,\&\, (u)(y\,E\,u \equiv z\,E\,u))$	[2.5]
d	$\supset zZ \,\&\, (u)(x\,E\,u \equiv z\,E\,u)$	[*b*, *c*]
e	$\supset z = x$	]A2.2]
2.8	$x = y \,\&\, y = z \supset x = z$	[*c*, 2.7]

This is principle **T**.

a	$xZ \,\&\, (u)(x\,E\,u \equiv y\,E\,u) \supset x = y$	[A2.2}
2.9	$xZ \supset ((u)(x\,E\,u \equiv y\,E\,u) \supset x = y)$	
2.10	$xZ \supset ((u)(x\,E\,u \equiv y\,E\,u) \equiv x = y)$	[2.9, 2.5]

Given the zulogical condition, this combines 2.5, the indiscernibiltiy of identicals with 2.9, the identity of indiscernibles.

2.5 and 2.6 capture part of the drift of identity. Given that Wellington is The Iron Duke, whatever is said about Wellington is *ipso facto* said about The Iron Duke. (I refrain from saying, as most workers do, that whatever is said about *one* is said about *the other*, because there is no other; there is just the one, called by various names.) 2.9 is stronger: given that whatever is properly say*able* about Wellington is properly say*able* about The Iron Duke, and vice versa, Wellington *is* The Iron Duke and vice versa. Of course, it is not possible to say everything sayable about anything, so 2.9 does not directly provide a means of establishing material identity. But this matter will be treated again when the fundamentals of SRD are presented.

Standard LPC, with '$x = x$' detachable, provides '$(\exists u)(u = u)$'. Whether this unspecified thing that is identical with itself is the same unspecified thing such that it is *f* or ~ it is *f*, is a matter that LPC does not decide. But worse, it provides '$(\exists u)(u = x)$', which goes beyond a mere *je ne sais pas* or two and threatens to open the floodgates to '$(\exists u)(u =$ Pegasus)' and '$(\exists u)(u =$ abcd)' and endless gibberish. To avoid this seeming ontological catastrophe it must resort to censorship of notation. But the catastrophe is only imaginary. As shown above, '∃' has no formal zulogical import, only intuitive, ontologically ineffectual, "existential" import.

Definition

A definition is a statement of identity for an attribute. It is not for two attributes, since only one is involved. But two referential expressions are used, one a brief name and the other to give a full specification of the relevant attribute. In actual cases, with constant expressions, defined attributes zu. But in this exposition, blankmarkers are used for generality. If SRD identity were called on for definitions then any definitional identity would warrant a zulogical claim, via 2.3, even if the blank were filled with garbage. What is needed is a modification of the identity relation that lacks zulogical commitment. The conventional relation suits the purpose, giving the same results as SRD identity except that zulogy is disregarded. A new notation might be used, but to avoid that I keep the same notation but put it in the scope of an operator '*D*'. This is in general prefixed to a detached formula and read "dependently". In a definition, it is appended to the identifier and read "by definition". With abeyance of zulogy, the definitional analogue of 2.10 needs no '$x\ Z$' and, because of 'D', A1.1 is not applicable.

A2.3 $D(x = y) \equiv (u)(x\ E\ u \equiv y\ E\ u)$

2.11 $x = y \supset D(x = y)$ [A2.3, A2.2]

The converse of 2.11 is not a theorem. Standard R, S, T and I are applicable. The form of a definition is D.

D $RE = [\phi(\text{it})]$

by definition, *RE* is (identical with) ...

In place of *RE* goes a novel referential expression which the definition dedicates to referring to whatever attribute is specified by the abstract.

a $\varphi(RE) \equiv \varphi[\phi(\text{it})]$ [D, 2.6]

b $x\ E\ RE \equiv x\ E\ [\phi(\text{it})]$

2.12 $x\ E\ RE \equiv x\ Z\ \&\ \phi(x)$

2.13 $x\ Z \supset (x\ E\ RE \equiv \phi(x))$

2.14 $x\ E\ RE \supset \phi(x)$

The first definition here repeats one given in *Definition*, ... It introduces 'B' for 'being', in the same sense effective in 'being wise is a virtue'.

2.1D $B f = [\text{it is } f]$

2.15 $x\ E\ B f \equiv x \text{ is } f$ [A1.1]

Whereas in *Definition*, ..., using variables, a further definition was given, now, using blankmarkers, I can derive a theorem from 2.1D.

2.16 $D(B\ r\ y = [\text{it } r\ y])$

2.17 $x\ E\ B\ r\ y \equiv x\ r\ y$

The theorems show that '$E\ B$' is eliminable, though 'B' without 'E' is not. It is not provable, in all generality, that $B\ f\ Z$ or that $B\ r\ y\ Z$.

Empty or garbage-filled forms of definition carry no zulogical commitment for the ostensible definiens or definiendum. Commitment attained for the definiens is transmitted to the definiendum, as shown in the next two chapters, to let SRD deal with talk about attributes, not just deal with attributes. Thus definitions can be effected indefinitely as long as what amounts to a thread of intelligibility is not broken.

So far, SRD can use the identity notation but not one for being identical with so-and-so. But now a generic identity atrribute is definable without circularity.

2.2D $B = y = [\text{it} = y]$

2.2D cannot be independently proved. The definition is stronger than the corresponding axiom, and the greater strength has a use for a familiar important purpose that Logic has not yet considered, much less catered for. This will be shown in a later volume.

BOOLEAN COMPOUNDS

A Boolean compound is a conjunction or an alternation or a negation of a conjunction or an alternation or a negation of a conjunction ... (to any depth) of intensions of a single kind. Primary Boolean compounds are compounds of simple attributes but there are derived Boolean compounds of other kinds of intensions: relations and predications. Sets (e.g. of dominoes or coins), aggregates or collections, not being intensions, are not subject to conjunction, alternation or negation. The association between simple and compound attributes is not a direct equivalence engendered by a mere notational abbreviation but a definition providing a conditional equivalence, the condition being zulogical.

Such definitions and conditional equivalences permit introduction, development, expression and discussion of intensions in disregard of any subject matter in which they are exemplified. Gibberish is prevented by the conditions, which restrict detached formulas and sentences to such as are meaningful. Therefore a system of language and logic is available in which the meanings of expressions do not depend in any way upon the how the world is. Judgments of truth or falsity of empirical propositions are dependent on available empirical propositions. Therefore meanings are not dependent on the truth of falsity of propositions.

In SRD, as in ordinary talk, meaning comes before truth.

Thus, although SRD, like other systems, begins with unanalysed sentences, then moves to predicate calculus, it does not follow most semantic efforts and base treatment of attributes on truth-functional aspects of sentences. In this volume I deal with attributes and relations, but predications and propositions are deferred until Volume 4, where truth is defined *simply and obviously*, without discussion of sentences.

Boolean attributes

The following abbreviations provide no logical advantage.

$(\sim f)x \equiv \sim(fx)$

$(f \,\&\, g)x \equiv (fx \,\&\, gx)$

$(f \vee g)x \equiv (fx \vee gx)$

Instead, by definitions, I provide characteristic connectives for the various intensions, using referential expressions.

3.1D $B\,n\,f = [B\,f\,Z\,\&\sim$ it is $f]$
being not f

3.1 x is $n\,f \equiv x\,Z\,\&\,B\,f\,Z\,\&\ \sim x$ is f

3.2 $x\,Z\,\&\,B\,f\,Z \supset x$ is $n\,f \equiv \sim x$ is $f)$

There cannot be *unconditional* equivalence between an affirmation and a negation. Introduction of the conventional "existential" quantifier flouts this principle of SRD.

3.3 Russell is a philosopher

3.4 Russell Z

3.5 ~ Russell is Greek

By 3.2, since Russell Z

If being Greek zu then Russell in *n* Greek

But consider the paradigm of a vacuous name, 'Pegasus'. To say that 'Pegasus' has no referent is to say that

~ Pegasus Z.

By 1.2,

~ Pegasus is *f*
~ Pegasus is *n f*
~ Pegasus is Greek
~ Pegasus is not Greek
~ Pegasus is dremple
~ Pegasus is !+aB2?
~ Pegasus = Pegasus

There is no way to get 'Pegasus' outside the scope of a negation when making an assertion, without having to retract '~ Pegasus *Z*'. Yet any trivial true remark about *x* suffices, by A1, to show that *x* zu.

If there is no such attribute as *B f* then there is no such thing to have a negation. If there is one, then clearly *B n f* is its negation.

3.2D *B neg B f* = [it = *B n f*]
being a negation of being *f*

a *x neg B f* ≡ *x Z* & *x* = *B n f*

b *B n f neg B f* ≡ *B n f Z* & *B n f* = *B n f*

3.6 *B n f neg B f* ≡ *B n f Z*

The treatment of conjunctions *B f* . *g* (being *f* and *g*) and alternations *B f* : *g* (being *f* or *g*) is along the same lines. In all cases, conditions are made for anything to be a Boolean. When the conditions are fulfilled, the defined thing is called such a Boolean.

3.3D *B f* . *g* = [it is *f* & it is *g*]
being *f* and *g*

3.7 *x* is *f* . *g* ≡ *x* is *f* & *x* is *g*

3.4D *B conj B f* ' *B g* = [it = *B f* . *g*]
being a conjunction of *B f* with *B g*

3.8 *B f* . *g conj B f* ' *B g* ≡ *B f* . *g Z*

3.5D *B f* : *g* = [*B f Z* & *B g Z* & it is *f* v it is *g*]
being *f* or *g*

3.9 *B f Z* & *B g Z* ⊃ (*x* is *f* : *g* ≡ *x* is *f* v *x* is *g*)

3.6D *B alt B f* ' *B g* = [it = *B f* : *g*]
being an alternation of *B f* with *B g*

3.10 *B f* : *g alt B f* ' *B g* ≡ *B f* : *g Z*

Boolean relations

A role of negation signs is to suspend zulogical commitment; this point was applied in ruling out "existential" quantification. '~ *x* is *f*', indeed '~ *p*', carries commitment for nothing. '*x* is *n f*' needs '*x Z*' and carries commitment for *x*. By substitution in 3.2,

3.11 $B\,n\,r\,y\,Z \,\&\, x\,Z \supset (x\,n\,r\,y \equiv {\sim}\,x\,r\,y)$

'*x n r y*' needs '*B n r y Z*' and '*x Z*' and carries commitment for *B n r y* and *x*. It does not need '*y Z*' and does not carry commitment for *y*. The requirement for a negated attribute is sufficient for an equivalent negation omitting the object(s) from its scope. The new negation sign 'ν' (Greek nu) is pronounced "not". No parentheses are needed since the different characters determine their different scopes.

3.7D $B\,\nu\,r\,y \;=\; B\,n\,r\,y$
being not *r* of *y*

3.12 $x\,\nu\,r\,y \equiv x\,n\,r\,y$

3.13 $B\,n\,r\,y\,Z \,\&\, x\,Z \supset (x\,\nu\,r\,y \equiv {\sim}\,x\,r\,y)$ [3.11]

3.14 $B\,n\,r\,y\; neg\; B\,r\,y \equiv B\,n\,r\,y\,Z$ [3.6]

3.8D $B\; R\text{-}neg\; B\,r\,y \;=\; [\text{it} = B\,\nu\,r\,y]$
being a relational negation of *y*

a $x\; R\text{-}neg\; B\,r\,y \equiv x = B\,\nu\,r\,y$

3.15 $B\,\nu\,r\,y\; R\text{-}neg\; B\,r\,y \equiv B\,\nu\,r\,y\,Z$

Other compound relations are more restrictive than the corresponding simple attributes, requiring a common object. By 3.8,

3.16 $B\,r\,y\,.\,s\,z\; conj\; B\,r\,y\,\text{'}\,B\,s\,z \equiv B\,r\,y\,.\,s\,z\,Z$

Here, reference is made to being *r* of *y* and *s* of *z*, with predicative conjunction. But relational conjunction and alternation involve conjoined relations with a common object. Ordinary English usage uses the same 'not' for both cases.

3.9D $B\,r\,..\,s\,y \;=\; B\,r\,y\,.\,s\,y$
being *r* and *s* of *y*

3.17 $x\,r\,..\,s\,y \equiv x\,r\,y\,.\,s\,y$

3.10D $B\; R\text{-}conj\; B\,r\,y\,\text{'}\,B\,s\,y \;=\; [\text{it} = B\,r\,..\,s\,y]$
being a relational conjunction of *B r y* with *B s y*

3.18 $B\,r\,..\,s\,y\; R\text{-}conj\; B\,r\,y\,\text{'}\,B\,s\,y \equiv B\,r\,..\,s\,y\,Z$

Alternation is analogous.

3.11D $B\,r\,::\,s\,y \;=\; B\,r\,y : s\,y$
being *r* or *s* of *y*

3.19 $x\,r\,::\,s\,y \equiv x\,r\,y : s\,y$

3.12D $B\; R\text{-}alt\; B\,r\,y\,\text{'}\,B\,s\,y \;=\; [\text{it} = B\,r\,::\,s\,y)]$
being a relational alternation of *B r y* with *B s y*

3.20 $B\,r\,::\,s\,y\; R\text{-}alt\; B\,r\,y\,\text{'}\,B\,s\,y \equiv B\,r\,::\,s\,y\,Z$

The proliferation of novel connectives here is necessary for making all the distinctions that *are there to be made*. Without them, similar connectives would have to be used in different contexts, with different scopes indicated by bracketing, a not strictly rigorous procedure. As shown in other contexts, brackets are best kept to a minimum. Bracketing is not a feature of spoken natural languages, though occasionally needed in their written forms, in cases of ambiguous scope. But a user need not be bothered by the new connectives. A user, as always, will use the ordinary English words and treat them with common sense when that suffices. Though these are ambiguous, users interpret them according to context in simple cases. The heavy machinery displayed here is merely for the purpose of regimenting ordinary usage in complicated cases, when intuitive skill is stretched too far.

Boolean theorems

Since 3.2, 3.7 and 3.9 for attributes and 3.11, 3.13, 3.17 and 3.19 for relations are iterative by blank filling (substitution), sentential theorems automatically provide other theorems.

3.21 If '*P*' '*F*', and '*R*' are replaced by respectively isomorphic sentential, predicative and relational notations and this

P

expresses a theorem then these

$B\,F\,Z\,\&\,x\,Z \supset x \text{ is } F$

$B\,R\,y\,Z\,\&\,x\,Z \supset x\,R\,y$

express theorems.

3.22 If '*P*' '*F*', and '*R*' are replaced by respectively isomorphic sentential, predicative and relational notations, and '*Q*' '*G*', '*S*' are treated likewise and this

$\sim P \vee Q$

expresses a theorem then these

$B\,F\,Z\,\&\,B\,G\,Z\,\&\,x\,Z \supset (x \text{ is } n\,F : G)$

$B\,F\,Z\,\&\,B\,G\,Z\,\&\,x\,Z \supset (x \vee R :: S\,y)$

express theorems.

This account does three things. 1) It draws attention to certain similarities among propositions, attributes in general and relations. 2) It provides different grammatical notations for the three cases, permitting discourse concurrently involved with the three kinds of intensions. 3) It formally makes clear certain zulogical conditions for deriving theorems, thus drawing attention to differences among the three cases. So far, there has been no *formal* reference to intensions, but in the next chapter I provide it for attributes, and in Volume 4 I do so for predications and propositions, by means of definitions of the various intensions.

The standard mathematical treatment of a "Boolean Algebra" provides none of the three things. An expositor may discuss propositions, "properties", sets and relations in a (commonly ungrammatical) vernacular context. Then he specifies a single neutral notation for an uninterpreted calculus and leaves a user to find his own interpretation. In effect, the multi-purpose conjunctions of ordinary langauge, with all their ambiguities, have been adopted, but in a region where common sense is not allowed to operate. Resulting discourse concerns either propositions or "properties" or sets or relations exclusively, never jointly as in, e.g., a consequence of 3.19, where '~', '*n*', and 'v' are signs of different sorts of negation, and 'v', ':' and '::' are signs of different sorts of alternation. These would be respectively conflated into a single negation sign and a single alternation sign in a single interpretation of a "Boolean Algebra" needing illicit intuitive bracketing to specify the scopes that the new signs have by definition.

$$\sim x\ r :: \vee\ s\ y\ \vee\ x\ r\ y : n\ s\ y$$

$$\sim (x\ r \cup \sim (s)\ y)\ \cup\ (x\ r\ y \cup \sim (s\ y))$$

A steep and thorny way

This illustrates a fundamental feature of the traditional mathematical method: excessive concentration on generality, with insufficient attention to specific differences. The prime example of such generality lies in the profusion of disparate entities treated in the supposedly single category "number". A user needs intuition and experience in applying a procedure, and the room for misapplication is great. In SRD, such specific differences are exposed by "hairsplitting", to provide rules governing correct application as well as correct development of a calculus. The case of number is treated in Volume 3.

Call on blankmarkers and zulogy instead of "variables" and censorship is necessitating a steep and thorny way, still not completed, instead of the primrose path trodden by conventional workers. Booleans are not being handed to me on a plate, for pronouncements about them are thickly hedged with zulogical requirements. I cannot yet categorically call anything a Boolean, but by the same token, pseudo-Booleans are not forced on me, because the conditional identity of definitions carries no zulogical commitment, nor does logical equivalence.

This chapter shows that zulogy is concerned with *referential integrity*, achieved without the futile gestures of existential axioms yet without call on intuitive global censorship of notations.

ATTRIBUTES

Background and aim

The trend of the exposition so far has been to develop versatile means of expression, regarding intensions, whose abuse is prevented by zulogical conditions. But the conditions presently prohibit intelligible discourse along with gibberish. It is good that I am not forced to accept being pqrst as an attribute, but bad that I cannot acknowledge such an obvious candidate as being blue. So it appears that philosophical neutrality is ensured by philosophical vacuity.

There is another problem. This paper has followed universal practice in using the vernacular quite freely in discussing intensions. But the calculus itself is like *Hamlet* without the Prince of Denmark, for so far nothing has been *formally* described as an attribute. This matters here, for the present work does not seek to avoid philosophy altogether (as if that could be done) nor to convince by philosophizing in ordinary language and exhibiting borrowed notations as pedagogical examples supplementary to the main argument. Rather, only formal plain-language statements here constitute philosophizing; the informal plain-language context is a mere guide to motivation and technique, and may be complete expunged if it fails in that role. All my informal philosophizing was done in the preface.

A method is needed of selectivly breaking the zulogical barrier by affirming, in some particular case, that *B f* zu. What is an attribute? Being a unicorn, being a rider of Pegasus and being a parent of Oliver Twist are intelligible, and that suffices here to make them attributes. It is not required that any unicorn zu or that Pegasus zu or that Oliver Twist zu, and the calculus neither presupposes nor implies these. But what is the root of intelligibilty? I do not dogmatize. Being a unicorn (a horned horse) is intelligible because being a horse is intelligible and being horned is intelligible. Why? because there *are* horses and there *are* horned things. If I take something exemplified as an attribute, I anchor discussion of intensions in subject matter — any user's subject matter. Users will call on concrete subject matter located in the physical world. I shall call mainly on abstract subject matter, about which it is silly to ask *where* it is.

Means

Now I come to the technicalities. It is fruitless to attempt to define being an attribute since application of being an attribute to being *f* requires that *B f Z*.

***D** *B* an *att* = [ϕ(it)]

B f is an *att* ≡ *B f Z* & ϕ(*B f*)

Without suitable cases where *B f Z* my problem here is to find cases where *B f* is an *att* **so that** *B f Z*. Hence I must use an axiom, making '*att*' a primitive notation. The specification of *B* an *att* must not always involve '*B f*' occurring before 'is' since I can have nothing to say about any attribute until I find one. But I can let '*B f*' occur after '*E*' because of the condition that, for a primitive non-compound attribute, something exemplifies it. Because of the feature discussed above regarding Oliver Twist, that intelligible use of 'a parent of Oliver Twist' does not demand that Oliver Twist zu, intelligible use of 'an exemplifier of *B f*' does not demand that *B f* zu; the primitive notation '*E*' carries the requisite meaning. Therefore the generic relational predicate '*E x*' can be used in an axiom, leaving it to be determined that in certain cases *x* is an *att*, so that *hence*, not as a presupposition, *x* zu. The axiom will clarify this perhaps difficult point. Note the calls here on dependent identity.

Specification

A4.1 (1) $y\ E\ B\ f\ \&\ D(x = B\ f)$

(2) v $B\ r\ u\ Z\ \&\ B = y\ Z\ \&\ D(x = B\ r\ y)$

(3) v $B\ f\ Z\ \&\ D(x = B\ n\ f)$

(4) v $B\ f\ Z\ \&\ B\ g\ Z\ \&\ D(x = B\ f\ .\ g)$

(5) v $B\ f\ Z\ \&\ B\ g\ Z\ \&\ D(x = B\ f : g)$

(6) v [see later volume]

$\supset$ *x* is an *att*

Compound attributes

(3) to (5) provide Boolean compounds.

4.1 $B\ f\ Z\ \&\ D(B\ n\ f = B\ n\ f) \supset B\ n\ f$ is an *att*

By elimination of the dependent identity,

4.2 $B\ f\ Z \supset B\ n\ f$ is *att*

4.3 $B\ f\ Z\ \&\ B\ g\ Z \supset B\ f\ .\ g$ is an *att*

4.4 $B\ f\ Z\ \&\ B\ g\ Z \supset B\ f : g$ is an *att*

These theorems provide Booleans, given components. They reduce the zulogical requirements for compound attributes in 3.6, 3.8 and 3.10 to the relevant component attributes.

Relations

By (2),

4.5 $B\ r\ u\ Z\ \&\ B = y\ Z \supset B\ r\ y$ is an *att*

This shows that the calculus is being successfully made intensional. That being married to Pegasus is an attribute does not presuppose or imply that Pegasus *Z*; it suffices that being Pegasus is intelligible, is an attribute. In conjunction with the previous iterative theorems 4.5 iterates to certify all Boolean compounds of relational attributes. Note here how the various connectives obviate the need for parentheses.

4.6 $B\ r\ u\ Z\ \&\ B\ s\ z\ Z\ \&\ B = y\ Z \supset$
$B\ r\ y$ is an *att* & $B\ n\ s\ y$ is an *att* & $B \vee s\ y$ is an *att*

4.7 $B\ r\ u\ Z\ \&\ B\ s\ z\ Z\ \&\ B = y\ Z \supset$
$B\ r\ y\ .\ n\ s\ y$ is an *att* & $B\ r\ ..\ \vee\ s\ y$ is an *att* &
$B\ r\ y : s\ y$ is an *att* & $B\ r\ :: s\ y$ is an *att*

These theorems reduce the zulogical requirements in 3.15, 3.18 and 3.20.

Predicability

By (1),

4.8 $y\ E\ B f\ \&\ D(B f = B f)\ \supset\ x$ is an *att*

4.9 $y\ E\ B f\ \supset\ B f$ is an *att*

4.10 y is $f\ \supset\ B f$ is an *att*

4.11 y is $f\ \supset\ B f Z$

Although '*y r z*' does not in general provide or require '*z Z*', for the particular relational predicate, '*E*', '*y E z*' provides '*z* is an *att*', hence provides, without requiring, '*z Z*'. This is a result of A4.1 that a corresponding definition would not supply.

A similar contrast occurs with the next definition.

4.1D *B predl y* = [*y E* it]
being predicable of *y*

4.12 *Bf predl y* ≡ *B f Z* & *y* is *f*

'*B f Z*' is required here by A2.1. By 4.11,

4.13 *Bf predl y* ≡ *y* is *f*

Without A4.1, so without 4.11, '*B f Z*' would still be unobtainable so 4.1D would be ineffective. Thus are definitions zulogically non-creative.

4.1D requires mention of some determinate constituent of the universe by use of some replacement for '*y*'. The next definition requires some unspecified constituent.

4.2D *B predl* = [(#*u*) *u E* it]
being predicable

4.14 *Bf* is *predl* ≡ *B f Z* & (#*u*) *u is f*

4.15 *x* is *f* ⊃ *Bf* is *predl* [4.11, 1.4]

As explained in *Definition*, ..., an abstract is available for a reflexive attribute without the generation of Russell's antinomy.

4.16 *x E* [it *E* it] ≡ *x E x*

4.17 *x E* [~ it *E* it] ≡ *x Z* & ~ *x E x*

4.18 [~ it *E* it] *E* [~[~ it *E* it] *E* [~ it *E* it]] ≡ [~ it *E* it] *Z* & ~ [~ it *E* it] *E* [~ it *E* it]

Without R2.2, for replacing 'it' along with '*x*', the antinomy arises.

* [~ it *E* it] *E* [~ it *E* it] ≡ [~ it *E* it] *Z* & ~ [~ it *E* it] *E* [~ it *E* it]

Definition, ... shows that use of a "bound variable" instead of 'it' or introduction of a putative name such as 'being impredself' produces the antinomy, since each prevents the needed substitution of the abstract "inside itself".

The source of attributes

A4.1 has taken me as far as it can but still the goal of pronouncing something an attribute eludes me. Discussion of compound attributes (e.g. in 3.8) requires compound attributes that can be discussed; compounds are available for discussion if their component attributes are; primitive non-compound attributes are presented by their exemplars in subject matter. By 4.10, if *x* is *f* and *y* is *g* and ... then *B f* is *att* and *B g* is *att* and ... so all Boolean compounds of these are attributes. Although some attributes are unexemplified they cannot all be for in the absence of describable items (subject matter) no attributes are given — and no one can create any.

A user finds attributes in his subject matter and it would be presumptuous for me to profess to lay bounds on the range of subject matter to which an investigator may find access. Given primitive exemplified attributes a user may discuss compounds.

Snowdon is a mountain

B a mountain is an *att*

(#*u*)(*u* is a wedding ring and *u* is golden)

B golden is an *att*

B golden . a mountain is an *att*

The absence of any known golden mountain in no way debars the conclusion that *B* golden . a mountain *Z*. By analogous argument *B* round . square is *att*.

Abraham is a father of Isaac

B a father of Isaac is an *att*

B = Pegasus *Z* ⊃ *B* a father of Pegasus is an *att*

B = Pegasus *Z* ⊃ *B* golden . a mountain : v a father of Pegasus is an *att*

Logical attributes

Still the full aim is not achieved. Attributes so far acknowledged have been concerned with particular subject matter. But a philosophically worthwhile calculus must provide discussion of attributes that in no way depend on particular subject matter. Such discussion must take the form of theorems, derived solely from axioms and definitions without reliance on judgments made in respect of particular subject matter. The machinery for development is to hand but no primitive attributes are available. Primitive attributes are presented by subject matter so there is need of at least one constituent of the universe that transcends all particular subject matter, to be an exemplar of non-compound logical attributes. As I have shown, J is such an item, temporarily taken on trust but inextricably involved with the calculus itself, apart from all applications. It is for the time being left to the reader to guess, if interest is sufficient, what J is and how it comes to play such a role. Let it just be said that no one can have any rational grounds for denying that J does what SRD claims for it.

By the temporary axiom:

4.19 J = J [TA, 2.3]

4.20 *B* = J is *predl*

4.21 *B* = J is an *att*

Finally I have a *theorem* acknowledging an attribute. The foundations of SRD are now complete and development may proceed. But zulogy still guards against chimaeras.

DEVELOPMENT

Universal and null attributes

Next I acknowledge some attributes that everything exemplifies and some that nothing exemplifies, and then I characterize such attributes.

5.1D *B idsom* = [(#*u*) it = *u*]
being (identical with) something

5.1 *x* is *idsom* ≡ *x Z* & (#*u*) *x* = *u*

a *x Z* ⊃ *x Z* & *x* = *x*

b ⊃ *x Z* & (#*u*) *x* = *u*

c ⊃ *x* is *idsom*

d *x* is *idsom* ⊃ *x Z*

5.2 *x* is *idsom* ≡ *x Z* ≡ *x* = *x*

5.2D *B Arist* = [(*u*)(it *E u* ∨ ~ it *E u*)]
being Aristotelian

5.3 *x* is *Arist* ≡ *x Z* & (*u*)(*x E u* ∨ ~ *x E u*) ≡ *x Z*

5.4 *x* is *Arist* ≡ *x* is *idsom* ≡ *x E* [it *Z*]

5.5 J *E* [it *Z*]

5.6 J is *idsom*

5.7 J is *Arist*

5.8 [it *Z*] is an *att*

5.9 *B idsom* is an *att*

5.10 *B Arist* is an *att*

5.11 *B n idsom* is an *att*

5.3D *B idnoth* = *B n idsom*
being (identical with) nothing

5.12 *B idnoth* is an *att*

5.13 *x* is *idnoth* ≡ *x* is *n idsom*

a ≡ *x Z* & ~ *x* is *idsom*

b ≡ *x Z* & ~ *x Z*

5.14 (*x*)(~ *x* is *idnoth*)

I now characterize these attributes. As with conjunctions, etc., without this step, people either illicitly talk informally (and usually incoherently) of attributes or resort to metadiscussion of predicates or other notations. In all cases (*pace* Tarski), SRD can be developed to incorporate such "metadiscussion" in ordinary talk.

5.4D *B* a *univatt* = [(*u*)(~ *u Z* ∨ *u E* it)]
being a universal attribute

5.15 *x* is a *univatt* ≡ *x Z* & (*u*)(~ *u Z* ∨ *u E x*)

a $xZ \supset xE$ [it Z] & xEB *Arist* & xEB *idsom* [5.4]

5.16 $(u)(uZ \supset uE$ [it Z]) ... *E B Arist*) ... *E B idsom*) [R1.1]

5.17 [it Z] ... *B Arist* ... *B idsom* is a *univatt*

5.5D *B* a *nullatt* = [it is an *att* & $(u)(\sim uE$ it)]
being a null attribute

5.18 *x* is a *nullatt* ≡ *x* is an *att* & $(u)(\sim uEx)$

a *B idnoth* is an *att* & $(u)(\sim uEB$ *idnoth*) [5.12, 5.14]

5.19 *B idnoth* is a *nullatt*

Variety of attributes

Is it inevitable that if attribures are countenanced at all then attributes of attributes are countenanced, for merely categorizing something as an attribute is assigning an attribute to it. Already various attributes have been classified as universal or as null. In general, attributes seem to be as varied and idiosyncratic in their characteristics and interrelations as items of any subject matter are.

5.20 *B predl* is *predl*

5.21 *B predl* is an *att*

5.22 *B* an *att* is *predl*

5.23 *B* an *att* is an *att*

5.24 *B Arist* is *Arist*

5.25 *B* a *nullatt* is an *att*

5.26 *B* a *nullatt* is *predl*

5.27 *B* a *univatt* is *idsom*

Even this brief examination of attributes beings out some interesting points. There are "self-exemplifying" attributes, since *B* an *att* is an *att*, *B idsom* is *idsom* and *B predl* is *predl*. There are mutually predicable (and exemplifying) attributes, since *B predl* is an *att* and *B* an *att* is *predl*. There are extended closed chains of attributes, since *B predl* is *idsom*, *B idsom* is a *univatt*, *B* a *univatt* is an *att*, *B* an *att* is *Arist* and *B Arist* is *predl*. It is clear that, whatever the case for particular sorts of attributes, attributes are not, in general, stratified into levels or types. Consequently, any calculus that depends on a hierarchy of predicate notations will not be fruitfully interpretable for discussing attributes.

There is a zulogical asymmetry between a constituent of the universe and a corresponding identity attribute. Any constituent must have such an attribute but such an attribute need not have an exemplar.

5.28 $xZ \supset xEB = x$

a $xZ \supset B = x$ is *predl*

5.29 $xZ \supset B = xZ$

a $B = xZ \supset B = x = B = x$

5.30 $B = x\,Z \supset B = x\,Z$

It cannot be proved that $B = x\,Z \supset x\,Z$. So there is room for unexemplified identity attributes such as $B =$ *Pegasus* and $B =$ Oliver Twist.

Entailment

I here treat entailment in connection with attributes, leaving implication for analogous application to predications.

5.6D $B \Rrightarrow y$ = [it is *predl* & $(u)(\sim u\,E$ it $\vee\, u\,E\,y)$]
being an entailer of y

With the implicit copula, '⇛' in a formula may be read 'entails'.

a x is *predl* $\supset (x \Rrightarrow y \equiv (u)(\sim u\,E\,x \vee u\,E\,y))$

b $\sim u\,Z \supset \sim u\,E\,x$

c y is a *univatt* $\supset \sim u\,Z \vee u\,E\,y$

d $\supset \sim u\,E\,x \vee u\,E\,y$

5.31 y is a *univatt* $\supset (u)(\sim u\,E\,x \vee u\,E\,y)$

5.32 B *Arist* is a *univatt*

5.33 $(u)(u\,E\,x \supset u\,E\,B\,Arist)$

5.34 x is *predl* $\supset x \Rrightarrow B\,Arist$

The conditions that an entailer be predicable, not just an attribute rules out such an unwanted vacuous result as the following.

* x is a *nullatt* $\supset x \Rrightarrow y$

Entailment is obviously transitive, and extensions follow routinely.

a $x \Rrightarrow y \;\&\; y \Rrightarrow z \supset x$ is *predl*
$\& (u)(\sim u\,E\,x \vee u\,E\,y) \;\&\; (u)(\sim u\,E\,y \vee u\,E\,z)$

b $\& (u)(\sim u\,E\,x \vee u\,E\,z)$

5.35 $x \Rrightarrow y \;\&\; y \Rrightarrow z \supset x \Rrightarrow z$

5.7D $B \Leftrightarrow y$ = [it $\Rrightarrow y \;\&\; y \Rrightarrow$ it]
being a co-entailer of y

'⇔' may be read as 'co-entails'.

5.36 $x \Leftrightarrow y \equiv x \Rrightarrow y \;\&\; y \Rrightarrow x$

5.37 x is *predl* $\supset x \Leftrightarrow x$

5.38 $x \Leftrightarrow y \equiv y \Leftrightarrow x$

5.39 $x \Leftrightarrow y \;\&\; y \Leftrightarrow z \supset x \Leftrightarrow z$

5.40 B *idsom* $\Leftrightarrow B$ *Arist* $\Leftrightarrow$ [it Z]

5.8D $B \Rightarrow y$ = [it $\Rrightarrow y \;\&\; \sim y \Rrightarrow$ it]
being a proper entailer of y ('⇒' may be read as 'properly entails'.)

5.41 $x \Rightarrow y \equiv x \Rrightarrow y \;\&\; \sim y \Rrightarrow x$

5.42 $B =$ J $\Rightarrow B$ *idsom*

Neither of two mutually exclusive attributes entails the other.

5.9D B *mutex* y = [it is *predl* & y is *predl* & it $n \Rrightarrow y$ & y $n \Rrightarrow$ it]
being mutually exclusive with y

5.43 x *mutex* y ≡ x is *predl* & y is *predl* & x $n \Rrightarrow y$ & y $n \Rrightarrow x$

Entailment theorems follow directly from Boolean theorems authorized by 3.22, hence from sentential theorems.

If '*P*' '*F*', and '*R*' are replaced by respectively isomorphic sentential, predicative and relational notations, and 'Q' 'G', 'S' are treated likewise and this

$P \supset Q$

expresses a theorem then these

$B\ F$ is *predl* & $B\ G$ is an *att* $\supset B\ F \Rrightarrow B\ G$

$B\ R\ y$ is *predl* & $B\ S\ y$ is an *att* $\supset B\ R\ y \Rrightarrow B\ S\ y$

express theorems.

Here are some examples, reduced by elision of tautologies.

a $p \supset p \vee q$

5.44 $B f$ is *predl* & $B\ g$ is an *att* $\supset Bf \Rrightarrow Bf : g$

a $p\ \&\ q \supset p$

5.45 $B f . g$ is *predl* $\supset Bf . g \Rrightarrow B\ g$

5.46 $B\ r\ y$ is *predl* & $B\ s\ y$ is an *att* $\supset B\ r\ y \Rrightarrow B\ r :: s\ y$

5 47 $B\ r .. s\ y$ is *predl* $\supset B\ r .. s\ y \Rrightarrow B\ r\ y$

Note that entailment is not intrinsically concerned with meaning. *B* idsom, *B* Arist and [it *Z*] are distinct attributes but they are truth-functional equivalents. Being the victor at Waterloo is distinct from being the Iron Duke, without even any logical connection. It just happens that Arthur Wellesley, who won at Waterloo, later in life became called the Iron Duke. Other results, e.g., that being Pegasus is an attribute even though there is no Pegasus, stem from a fundamental polarity of the subject/predicate form. Frege's multi-argument form, with its ill advised obliteration of that polarity, cannot express, much less prove, these results. Again, and not for the last time, one sees that standard Logic has too few degrees of freedom. It is truly a Procrustean bed.

At last, relations

Finally I formalize reference to relations. Note the implicit use here of "existential" quantification, which has no zulogical commitment. This definition supports the case for being a rider of Pegasus as a relation.

5.10D B a *rel* = [~(u)~ ($B = u\ Z$ & it $= B\ r\ u$)]

a x is a *rel* ≡ ~(u)~ ($B = u\ Z$ & $x = B\ r\ u$)

5.48 $B = y\ Z$ & $B\ r\ y\ Z \supset B\ r\ y$ is a *rel*

5.49 $B\ r\ y$ is a *rel* ≡ $B\ r\ y$ is an *att*

TOWARDS THE VERNACULAR

The prime requirements of SRD are canonical intelligible interpretation and formal rigour. Given these, the closer the system is to everyday usage the better. A remarkably close approximation is possible if natural language is taken as a guide rather than rejected as a hopelessly imprecise instrument. What follows is only the beginning of a process that will stretch over many volumes, and need not stop as long as intellectually useful expressions are lacking in SRD.

Relative clauses

It is apparent that

A) x, which is f, is $g \equiv x$ is f & x is g

The LHS, containing a relative clause, seems to be a mere notational equivalent of the RHS, a simple conjunction. So far the relative clause notation offers no new power of expression.

B) $\sim x$, which is f, is $g \equiv \sim x$ is $f \vee \sim x$ is g

Here the LHS is the simple negation of the previous LHS. As long as only sentential negation is available, so that the LHS is negated as a whole, no advantage is taken of the novel structure of the LHS. The relative clause still contributes nothing and theoreticians have not been motivated to employ it. But

C) x, which is f, is not $g \equiv x$ is f & $\sim x$ is g

Here is no mere negation of either side of *A*). Although the *predication* that x is g has been negated the *presupposition* that x is f has been accepted. The reverse might have happened

D) x, which is not f, is $g \equiv \sim x$ is f & x is g

What is suggested above is trivial from a naive, ordinary-language stance. But in the sphere of precise discourse it becomes problematic because standard notation is not expressive enough to deal with it. The simple reason is that the naive presentation uses an unexplained predicate negation, and standard attempts at rigour cannot provide the required explanation, which calls on zulogy. Following the strategy of the present project, already practised in the treatment of attributes, I now formalize "object-language" discussion employing relative clauses and in volume 4 I shall develop *the same* language to make meta-judgments about presuppositions and such. SRD, like a natural language, has no need of a separate "meta-language".

6.1D $B = y\ \omega$ is f = [it $= y$ & y is f]
being (identical with) y, which is f

The notation for 'which', not phonetically but pictorially apt (in English), is a small omega.

6.1 $x = y\ \omega$ is $f \equiv x = y$ & y is f

6.2 $x = x\ \omega$ is $f \equiv x$ is f

a x is $f \supset xZ$ ... $\supset x\ \omega$ is fZ [6.2]

b	$x\ \omega\ \text{is}\ f = x\ \omega\ \text{is}\ f\ \equiv\ x\ \text{is}\ f$	[6.2]
6.3	$x\ \omega\ \text{is}\ f Z\ \equiv\ x\ \text{is}\ f$	[*a*, *b*]

Next *A*) is proved.

a	$x = x\ \omega\ \text{is}\ f\ \&\ x\ \text{is}\ g\ \supset\ x\ \omega\ \text{is}\ f\ \text{is}\ g\ \&\ x\ \omega\ \text{is}\ g\ \text{is}\ f$	[commutativity of '&']
b	$x\ \text{is}\ f\ \&\ x\ \text{is}\ g\ \supset\ x\ \omega\ \text{is}\ f\ \text{is}\ g$	
c	$x\ \omega\ \text{is}\ f\ \text{is}\ g\ \supset\ x\ \omega\ \text{is}\ f Z$	
d	$\supset\ x\ \text{is}\ f$	[6.3]
e	$x\ \omega\ \text{is}\ f\ \text{is}\ g\ \supset\ x\ \omega\ \text{is}\ g Z$	[*a*, 6.3]
f	$\supset\ x\ \text{is}\ g$	[6.2]
g	$x\ \omega\ \text{is}\ f\ \text{is}\ g\ \supset\ x\ \text{is}\ f\ \&\ x\ \text{is}\ g$	[*d*, *f*]
6.4	$x\ \omega\ \text{is}\ f\ \text{is}\ g\ \equiv\ x\ \text{is}\ f\ \&\ x\ \text{is}\ g$	[*b*, *g*] (for *A*))
6.5	$x\ \omega\ \text{is}\ f\ \text{is}\ g\ \equiv\ x\ \omega\ \text{is}\ g\ \text{is}\ f$	[6.4, commutativity of '&']

Then *B*) by sentential calculus.

6.6 $\sim x\ \omega\ \text{is}\ f\ \text{is}\ g\ \equiv\ \sim x\ \text{is}\ f \vee \sim x\ \text{is}\ g$ (for *B*)

Then *C*) and *D*) by substitution in 6.5:

6.7	$x\ \omega\ \text{is}\ f\ \text{is}\ n\ g\ \equiv\ x\ \text{is}\ f\ \&\ x\ \text{is}\ n\ g$	
6.8	$x\ \omega\ \text{is}\ f\ \text{is}\ n\ g\ \equiv\ x\ \text{is}\ f\ \&\ \sim x\ \text{is}\ g$	(for *C*)
6.9	$x\ \omega\ \text{is}\ n\ f\ \text{is}\ g\ \equiv\ x\ \text{is}\ n\ f\ \&\ x\ \text{is}\ g$	
6.10	$x\ \omega\ \text{is}\ n\ f\ \text{is}\ g\ \equiv\ \sim x\ \text{is}\ f\ \&\ x\ \text{is}\ g$	(for *D*)

It is not actually relative clauses that are specified by 6.1D. Only referential expressions for attributes are specifiable by definitions. If *y* exemplifies an attribute defined by a case of 6.1D, then is *y* ω is *f* zu. By substitution of qualified referential expressions for blankmarkers,

a	$x\ \omega\ \text{is}\ g = x\ \omega\ \text{is}\ g\ \omega\ \text{is}\ f\ \equiv\ x\ \omega\ \text{is}\ g\ \text{is}\ f$	[6.2]
6.11	$x\ \omega\ \text{is}\ g\ \omega\ \text{is}\ f Z\ \equiv\ x\ \text{is}\ g\ \&\ x\ \text{is}\ f$	[6.3, 6.4]
6.12	$x\ \omega\ \text{is}\ g\ \omega\ \text{is}\ f Z\ \equiv\ x\ \text{is}\ g\ .\ f$	
6.13	$x\ \omega\ \text{is}\ g\ \omega\ \text{is}\ f Z\ \equiv\ x\ \omega\ \text{is}\ g\ .\ f Z$	

A qualified referential exression may occur in object position.

6.14	$x\ \omega\ r\ y\ \omega\ \text{is}\ f Z\ \equiv\ x\ \omega\ r\ y\ \text{is}\ f\ \equiv\ x\ r\ y\ \&\ y\ \text{is}\ f$	[6.11, 6.4]
a	$y\ \text{is}\ f\ \supset\ (x\ r\ y\ \equiv\ x\ r\ y\ \omega\ \text{is}\ f)$	
6.15	$y\ \text{is}\ f\ \&\ x\ r\ y\ \supset\ x\ r\ y\ \omega\ \text{is}\ f$	[6.5]

Here, clearly, it is *y* that is *f*, by proximity of '*y*' to 'ω'. But substitution of '*ry*' for '*g*' (and elision of 'is') in 6.11 and 6.13 makes it the qualified *x*, and not *y*, that is *f*. Parentheses are needed here, as they are in ordinary language, for precision.

6.16 $(x\ \omega\ r\ y)\ \omega\ \text{is}\ f Z\ \equiv\ x\ \text{is}\ r\ y\ \&\ x\ \text{is}\ f$

6.17 $(x\ \omega\ r\ y)\ \omega\ \text{is}\ f Z\ \equiv\ x\ \omega\ r\ y\ .\ f Z$

All this is a laborious but necessary way to prove things that are intuitively obvious.

Restrictive clauses

As just shown, a non-restrictive relative clause, using 'which', is part of a referential phrase since its head is a proper name. Now I come to a restrictive relative clause, using 'that (is)'. Here, 'that' is part of a predicate phrase since its head is a common noun. Again I use a Greek letter as abbreviation. Proofs follow much as they do in the case of 'which'.

6.2D $B\,\mathrm{a}\,b\,\theta f\ =\ B\,\mathrm{a}\,b\,.\,f$
being a b that is f

6.18 x is a $b\,\theta f\ \equiv\ x$ is a b & x is f

6.19 x is a $b\ \supset\ (x$ is a $b\,\theta f \equiv x$ is $f)$

6.20 x is $f\ \supset\ (x$ is a $b\,\theta f \equiv x$ is a $b)$

6.21 x is a $b\ \supset\ (x$ is a $b\,\theta\, n f\ \equiv x$ is a $n f)$

6.22 x is n a $b\,\theta f\ \equiv\ x$ is n a $b \vee x$ is $n f$

It might seem that the new notation, being seemingly synonymous (because of the '=' used in definitions, with the old, can serve no useful purpose. Perhaps this rash judgment has led logicians to go their own ways rather than copy the vernacular. But in 'a *b*. *f*', 'a *b*' and '*f*' are predicative notations and permit interchange to '*f* . a *b*', whereas in '*b* θ *f*', '*b*' and '*f*' are of different syntactic kinds. Although, throughout a formula, '*b*' and '*f*', being mere blankmarkers, are interchangeable (in all occurrences), interpreted fillings are not. 'a man that is tall' is poles apart from 'a tall that is man'. The difference in syntactic structure between the two sides of 6.2D gives the vernacular, and now the present calculus, great expressive power.

It is commonplace nowadays to recognize the merely quasi-referential role of 'a man' in 'Russell met a man.'

Russell is a man	(vernacular)
Russell is a man	(calculus)
Russell met a man	(vernacular)
(#*u*)(Russell met *u* & *u* is a man)	(calculus)

The new notation permits substitution of '*b* θ *f*' for '*b*'.

Russell met a man that is tall

(#*u*)(Russell met *u* & *u* is a man θ tall)

(#*u*)(Russell met *u* & *u* is a man & *u* is tall)

In SRD, 'a man' is always predicative. Reference is performed by names or by quantificational apparatus. But, once more, ordinary language has a lesson that is well worth while heeding.

Brief quantificational reference

'A *b*', in origin a predicate expression, is often elliptically used referentially in ordinary talk.

(#u)(Russell met u & u is a man) (calculus)
Russell met a (i.e. some) man (vernacular)

This form of talk is so useful, in its brevity, and so *familiar*, (so familiar, in fact, that hardly anyone ever notices its peculiarity) that it warrants formalizing. I introduce a quasi-identity, working only from an ordinary reference to the new quantificational reference, along with a Greek letter 'σ' for 'some', and a rule of inference. The new notation can be pronounced 'is' in a formula, since the context overrides ambiguity.

6.3D $B \to \sigma\, b$ = [(#u)(u is a b & u = it]
being some (i.e. quasi-identically a) b

A6.1 $\phi(\sigma\, b) \equiv \phi(x)$ & $x \to \sigma\, b$

Whereas 'a b' is a predicate expression, 'σ b' is a referential one, which is why I talk about quasi-identity, not quasi-predication.

a $x \to \sigma\, b \equiv$ (#u)(u is a b & $u = x$) [6.3D]

x is a $b \supset x \to \sigma\, b$

Napoleon → σ man

Wellington → σ man

Russell met x & $x \to$ σ man ≡ (#u)(u is a man & $u = x$ & Russell met x)

Russell met σ man ≡ (#u)(u is a man & Russell met u) [A6.1]

In this example, a particular man is singled out. But the virtue of the brief expression becomes more evident when multiple non-identified *b*s are referred to, as in this, which will be developed further when text is discussed, in Volume 5.

D *B* a stg = *B* a char : → σ stg ^ σ stg
being *a* character or *some* string followed by *some* string

The implicit preposition in '^' requires the referential 'some'. D does explicitly and grammatically with the quantifier 'some' what the vernacular does elliptically with use of 'a' both as an article (with 'is') and as a quantifier (with '=').

[it is a char v (#u)(#v)(u is a stg & v is a stg & it = u ^ v)]

Quasi-identity is not reflexive, commutative or transitive. If it were, or if ordinary identity were called on, then invalid results like the following would ensue. Napoleon is some man. Some man is Wellington. Napoleon is Wellington.

Definite description and reference

Now I demonstrate even greater versatility for the *definite* article than for the indefinite. Treatment proceeds in four stages: 1) unique description, 2) definite description, 3) the import of a definite reference and 4) achievement of such reference. I follow with 5) treatment of an Olympian *fiat*, which "creates" a referent.

1) *Unique description* concerns an attribute that some unique exemplar exemplifies, *if any*. Unique description is *patently* concerned with an attribute and a predicate.

6.4D *B uqf* = [it is f & (u)(u is $f \equiv u$ = it)]
being uniquely f

6.23 x is $uq\,f \equiv x$ is f & $(u)(u$ is $f \equiv u = x)$

6.24 x is $f \supset (x$ is $uq\,f \equiv (u)(u$ is $f \equiv u = x))$

There is no zulogical problem if there is no exemplar or there are more than one exemplars; the attribute is merely null, since if either there is no *b* or there are more than one *b*s, nothing is *uniquely* a *b*.

a $\sim x$ is $uq\,f \equiv \sim x$ is $f \vee \sim(u)(u$ is $f \equiv u = x))$ [6.23]

b $\sim x$ is $uq\,f \equiv \sim x$ is $f \vee \sim(u)\sim(\sim u$ is f & $u = x \vee u$ is f & $\sim u = x))$

6.25 $\sim x$ is $f \supset \sim x$ is $uq\,f$

6.26 $\sim(u)\sim(\sim u$ is f & $u = x) \supset \sim x$ is $uq\,f$

6.27 $\sim(u)\sim(u$ is f & $\sim u = x) \supset \sim x$ is $uq\,f$

2) *Definite description* concerns an attribute and a predicate for unique description, though the use of a definite article can make this *unobvious*. In ordinary English, 'the' can take the place of 'uniquely a' as in 'Mt Everest is the (i.e. uniquely a) highest mountain'. Here 'is' is a copula and 'the highest mountain' is a predicate phrase.

6.5D *B t b* = *B uq* a *b*
being the *b*

Apart from 'the Deity' and 'the Universe' (and even here uniqueness is commonly flouted) there is little demand for 'the *b*' with a simple common noun. Usually there is some qualification, some restriction. That is what restrictive clauses are for.

6.28 *B t b th f* = *B uq* a *b th f*

6.29 x is $t\,b\,th\,f \equiv x$ is a $b\,.\,f$ & $(u)(u$ is a $b\,.\,f \equiv u = x)$ [6.23]

Notice the non-standard copula in place of the usual '=' here. The role of 'the', so far, is to indicate uniqueness of whatever (if anything) is referred to by suitable referential notation — it is *not* to indicate reference. The intent of the definition is shown by the definition of a *sortal attribute*, associated with a *copula*.

3) *The import of a definite reference* is determined by definition of a suitable attribute (only attributes can be defined). The import of the definition is shown by the definition of an *identity attribute*, calling on '=', not of a sortal attribute. *Definite reference* concerns, *by a speaker's intent*, a *sole exemplar* of some attribute. In ordinary English 'the' can take the place of 'something that is uniquely a' as in 'Mt Everest is the (i.e. is (identical with) something that is uniquely a) highest mountain. I use a Greek letter for this referential use of 'the'.

6.6D $B = \tau\,b$ = *B t b*
being (identical with) the *b*

Neither this definition nor use of it in an assertion determines that there is $\tau\,b$. Like any definition, it does not even determine that there is an attribute $B = \tau\,b$, since dependent identity, used in definitions, does not have zulogical import, for reasons explained earlier. No talk, not even talk making a definition or someone's claim to refer, can determine what zu. *Only the way things are can decide that.*

6.30 $x\,E\,B = \tau\,b \equiv x$ is $t\,b$ [A2.1]

a $x = \tau\,b \equiv x$ is $t\,b$

b x is $t\,b \supset \tau\,b = x$

c x is $t\,b \supset (\phi(\tau\,b) \equiv \phi(x))$ [2.6]

6.31 x is $t\,b \supset \tau\,b$ is $t\,b$ [*b*, 2.6]

Nothing here determines, as a theorem, that $\tau\,b\,Z$. Many people affirm that God zu, but many deny this. The matter cannot be settled by anyone's talk. Extra-logical evidence is needed.

Reference to the highest mountain is unproblematic because something *is* the highest mountain. But reference to the golden mountain is (according to available evidence and good scientific advice) impossible since there is no golden mountain to refer to. But you can mount an expedition to look for the golden mountain, and some people did, expecting it to be in Eldorado. If you are an entrepreneur you can plan how you will take possession of the golden mountain should one be found.

Sometimes one needs to talk hypothetically about the *b* without having determined whether there is one or not. Sometimes someone purports to talk about what you take to be a figment. No one has any difficulty in understanding what is said. SRD, too, by 6.31, is open to all kinds of non-committal talk, which requires only that the talk be intelligible, not that there be any suitable referents. But in *conventional mathematics*, mutually inconsistent assertoric axiomatic systems make the whole *formally inconsistent.* Formally, there both is and is not a choice set as specified by the axiom of choice. The crutch of an *informal* stipulation of hypotheticality does not really prevent inconsistency. It merely transfers the multiple activities to the realm of game-playing.

So far, the *b* can be mentioned given that some *named* object is *t b* (see 6.31). That is fine for a case like Mount Everest as the highest mountain on Earth. But the highest mountain on the Moon is not known and *has not been named*, since the far side has not been explored. On the strength of this philosophical (not geographical) discovery, I take the liberty of naming it in advance.

D B = Mount Trotter = B the highest mountain on the Moon

Nothing in SRD determines as a theorem that Mount Trotter zu, though a relevent identity attribute zu, just as nothing here determines that Pegasus zu, though a relevent identity attribute zu. But it is uncontroversial that there *is* the highest mountain on the Moon, that it zu. So *it happens that* Mount Trotter is (and, since the the following 'the' is ambiguous, is identical with) the highest mountain on the moon, and thus Mount Trotter zu. There is no (practical) doubt that Mount Trotter will eventually be discovered, and redundantly but justifiably and unwittingly, renamed. There is nothing novel in my "premature" naming. Riskier moves happen all the time when prospective parents, or even just trying parents, decide on a name for a not yet available baby.

But, as in ordinary language, it is not necessary in SRD to have a specific name for the *b*. It is immaterial what name is used, and there need not be any name at all. Due to the absence of the free blankmarker (variable) '*x*' from the RHS of 6.31.

6.32 $(\#u)(u \text{ is } t\, b) \supset \tau\, b \text{ is } t\, b$

The consequent here, expressed colloquially by the elliptical 'the *b* is the *b*', is not a tautology.

Sometimes a proper name for a thing is introduced before a corresponding sortal attribute has been introduced for things of a kind, so no public definite description can be given. I use a capital 'T' in this case, along with the initially capitalized name. The two prime characteristics of a proper name are that it has no (descriptive) meaning and that it refers to "you know what". If you don't know what then you have to ask somebody.

D $B = \mathrm{T\,B} \; = \; [\phi(\mathrm{it})]$
being The B

Here $[\phi(\mathrm{it})]$ can be anything that anyone offers to explain to someone else what the name refers to. It will be tailored for the particular recipient, with no guarantee that it will work for others, or even, at the first attempt, for the original recipient. The matter is a practical one, to be dealt with in a practical manner, and is of *no logical concern.* Any investigation made into the details of such transactions is an investigation in psycholinguistics. In philosophy, as an investigation of "individual essences", it is a pseudo-problem.

a $x = \mathrm{T\,B} \equiv x\, Z\, \&\, \phi(x)$

The intention is to introduce a proper name that has no meaning but just refers to whatever one understands it to refer to. Many names have, however, private connotations for various users. These too are of no logical concern. The absence of meaning for proper names is well indicated by such expressions as 'The Jabberwock', 'Pretty Pattie', (for a race horse) and 'The Old Bull And Bush', for a public house. The proper name will be empty if there is no thing that fits the offered description. Again, imagine consulting a large census database about John Smith. You do not know all the John Smiths and, if all the details are combined, the putative John Smith is someone with contradictory features: he is 23, 47 and 71 years old, 165, 178 and 190 cm tall, etc. You object: "But you have to know which John Smith is in question.". That is no objection; that is what I just said.

4) Now for *achievement of definite reference*. Only one condition is necessary and sufficient for achievement of reference *of any kind*: the availability of the thing purportedly referred to. Except as shown here in discussing intensions, such availability is not a matter of logic. A successful attempt to introduce the purported name 'Pegasus' as the name of a mythical beast provides an identity attribute but it fails to conjure up any such beast. If it provides a grasp of the conditions required for reference to Pegasus, discourse using the name becomes intelligible.

Disregard of the indexicality of proper names is only one of many fatal flaws in extant systems. Lacking a proper form of definition, they cannot explain meanings; lacking zulogy, they cannot make the distinctions treated here. Set theory, relying solely on names, cannot even express, much less explain, meanings, and cannot make *reliable* reference to *publicly* identified referents. It is restricted to the "you know what" situation, with no means of explaining what. A surprisingly large number of people, both expert and lay, think they "know what". But, as with St. Augustine and time, they "know" only so long as they are not required to explain. I examine set theory in detail in Volume 2, where I tell you what.

5) SRD permits assertions about attributes, including *identity attributes*. It is not forced to presuppose referents for identity attributes. It permits a distinction between a speaker's *intent* to refer and the *achievement* of reference. It has a layer of discourse that is simply meaningful, without any referential import at all.

But extant systems, lacking such discriminative power, can introduce names only in assertions (indeed theorems) about the purported referents, so in the absence of such referents such assertions are not merely a user's mistakes, but false theorems. This fault is in part easily avoided by the prohibition of empty names. But definite references (used with '='), paradoxically called definite descriptions (properly used with 'is'), are essential for the aims of users. These expressions, like definite references proper in SRD, indicate a user's *intent* at reference (such is part of their meaning.). But extant systems have no insulating layer of discourse of this kind. A referential expression can be introduced only *in use*, *presupposing* the availability of a referent; it thus has the character of a *fiat*, purporting to create something, not just introduce a potential name. The unsolved standard problem is to have access to the introduction of effective referential expressions yet to avoid unwarranted purported reference.

No doubt, there will be some who find all this "hair-splitting" excessively difficult, and unnecessary. But difficulty depends on where you stand. A layman sees the difficulty and is dismayed (***nearly four pages***, (and I am not yet finished with 'the')). But an expert knows of other treatments, ancient and modern, of the present matters, that not only are vastly more "difficult" and full of dubious metaphysical pronouncements or presuppositions, but do not even work. Instead of following SRD, one can go in the Meinongian direction to conclude that there *is* the golden mountain, indeed the round square, in some mysterious sense of 'is'. Or one can, with Russell, consider "incomplete symbols", or, with Frege, take "concepts" as "unsaturated", and not "objects". Or one can consider "possible worlds", which raise the problem of "rigid designators", arising from features absent from SRD: the necessity of identity, the detachability of '$x = x$' and the "existential" import of that formula. This does not exhaust the contortions resorted to in attempts to solve pseudo-problems that are created by the inherent lack of versatility of extant systems of logic. *Definition*, ... explained another.

The virtue of my "hair-splitting" is that I can proceed in minute steps, each attended with minimal novelty. The result of leaving such hairs unsplit is to make big jumps, in which many relevant considerations are unseen. Thus arise problems that are insoluble because errors are made in the invisible presuppositions. There is no such faulty metaphysics in SRD. It just claims that intensions can be talked about, a fact substantiated by everyone, every day, even in disregard of misplaced metaphysical scruples.

REVIEW AND PROSPECT

In this first volume, I have introduced zulogy and a rigorous treatment of attributes, and shed new light on standard predicate calculus and its techniques. Along the way, I have made stringent criticisms of many extant methods in logic and mathematics. Yet I have found *nothing wrong* with these methods, taken on their own terms. My innovations are strategic, setting a locus of operations that differs radically from conventional ones. I am enabled to deal with problems with comparatively ridiculous ease, because I rely on the powers of ordinary English. Since traditional methods are based on an unproductive strategy, at odds with ordinary language, the enormous output of ingenious tactical operations during this century has succeeded only in getting workers into a deeper, muddier hole. That strategy is now *intellectually bankrupt*.

I promised an interpretation of set theory. At a suitable level of exposition, I can say in one line of text what a set is, and prove it. Like a detective-novel author, I gradually introduce a cast of characters, with whom you can become familiar, perhaps with no suspicions. Then, towards the end of Volume 2, I just say "The butler dunnit.". But it is a long job to give you a detailed, precise account of the characters, with all the 'i's dotted and all the 't's crossed. The need for such preliminaries perhaps explains why the set-theory bubble has not been burst before now, despite wide-spread, determined intentions a hundred years ago, before set theory caught on, and sporadic ones since it became entrenched. But being preparation for a demolition is a minor, incidental feature of the exposition. Its more lasting value lies in its positive uses in reasoning, which are shown or hinted at along the way.

A reader can go Volume 2 and peek at the answer. But first, he will probably not understand it, even though the words it calls are on perfectly familiar, because different people have different and multiple understandings of the words, and I have selected single meanings, and have regimented use of the words. Second, even if he does understand, he may not grasp the catastrophic conseqences for 20th-century logic and foundations of mathematics.

Third, the whole enterprise has something of the nature of a puzzle. Both in my mentioned previous publications and in Volumes 1 and 2 I have laid clues and dropped hints regarding the nature of sets, this being the $64,000 intellectual question of the century. If you think this is an exaggeration, think again after seeing the answer, bearing in mind the tremendous influence of set theory on intellectual activity in the 20th century. Any self-respecting expert should avoid being told the answer before he has pondered, and offered his own or given up. And there is a reward for finding the answer yourself, with a clear idea of its consequences. Seeing it brings a realization with the force of Paul's vision on the road to Damascus. You will experience what I call the OMG reaction; you will clap your hand to your brow and shout "O my God!" as you see the matter in its awesome historical perspective.

You wouldn't want to miss that, would you? So be on the lookout for people who have read Volume 2 and are keen to tell you the answer. Don't let them, or they will spoil your fun. And when *you* find out, don't be a spoilsport.